Degraded Forest Rehabilitation and Sustainable Forest Management in the Asia-Pacific Region

Asia-Pacific Network for Sustainable Forest Management and Rehabilitation

China Forestry Publishing House

Nature Conservation Publishing Center • China Forestry Publishing House
Chief editors: Shen Lixin, C.T.S. Nair
Editors: Pan Yao, Wang Jun, Hu Chuyu
Staff editors: Xiao Jing, He Na, Liu Jialing

图书在版编目(CIP)数据

亚太地区退化森林恢复与可持续森林管理 =Degraded Forest Rehabilitation and Sustainable Forest Management in the Asia-Pacific Region：英文 / 亚太森林恢复与可持续管理组织编著. -- 北京：中国林业出版社，2016.4
ISBN 978-7-5038-8521-1

Ⅰ. ①亚… Ⅱ. ①亚… Ⅲ. ①森林植被－恢复－研究－亚太地区－英文 ②森林管理－研究－亚太地区－英文 Ⅳ. ①S718.54②S75

中国版本图书馆CIP数据核字(2016)第095217号

All rights reserved. Reproduction and dissemination of material in this information product for educational or other non-commercial purpose are authorized without any prior written permission from the copyright holder provided the source is fully acknowledged. Reproduction of material in this information product for resale or other commercial purposes is prohibited without written permission of the copyright holder. Applications for such permission should be addressed to: Information Officer, APFNet Secretariat, 6th Floor, Baoneng Center, 12 Futong Dongdajie, Wangjing Area, Chaoyang District, Beijing 100102, People's Republic of China, or by email to: info@apfnet.cn. Queries for hard copies can also be addressed to the address above.
© 2016 APFNet

First Published in the P.R. China in 2016 by
China Forestry Publishing House
No.7, Liuhaihutong, Xicheng District, Beijing 100009
Printed and bound in Beijing, China

Foreword

In order to address widespread forest degradation, the international community has made several significant commitments, including the Bonn Challenge to restore 150 million hectares and the Asia-Pacific Economic Cooperation (APEC) 2020 Goal to boost forest cover by 20 million hectares by 2020. These commitments help contribute to mitigating levels of forest degradation across the globe.

However, the scale of the problem requires more action, especially given the increasing demand for food, fuel and fibre, and the need to strengthen a range of ecosystem services including maintenance and improvement of watersheds, improvement of carbon sequestration and storage, and protection of biological diversity.

Recognizing these concerns, APFNet organized the Workshop on Degraded Forest Rehabilitation and Sustainable Forest Management from1 to 12 July 2014 in Kunming, Yunnan Province, China. The workshop gathered together 15 forestry practitioners involved in degraded forest rehabilitation from 12 developing economies in Asia and the Pacific, to share experiences and best practices.

To better disseminate the output of the workshop and share more information about forest rehabilitation effort in individual economies, Degraded Forest Rehabilitation and Sustainable Forest Management in the Asia-Pacific Region, a compilation of the participants' reports, assesses the current state of forest degradation and rehabilitation efforts in each economy, and identifies future actions, in particular by drawing upon successes and failures in adopting ecosystem approaches for rehabilitation.

Some of the reports examine policy, planning and management aspects of the rehabilitation efforts in their economy, while others introduce experiences from projects and case studies. A number of the reports also give suggestions on how to improve on past rehabilitation efforts, which readers can adapt to their economies' specific contexts. We believe that they are immensely valuable in providing a reference point for degraded forest rehabilitation in the Asia-Pacific region.

We take this opportunity to express our appreciation and gratitude to the participants of the workshop who submitted their reports, and also to Dr. C.T.S. Nair and Professor Shen Lixin for compiling and editing the reports. We thank them all for their hard work and extensive contributions in making the compilation a success.

APFNet Executive Director

Table of Contents

Foreword

An Overview of Rehabilitation of Degraded Forest Lands in Asia
    C.T.S. Nair, Shen Lixin, Pan Yao ·· 001

Participatory Forest Management in Degraded Forests: Perspective REDD+ in Bangladesh
    Al-Amin, Hasan ·· 013

Forest Restoration and Plantations in Cambodia
    Piseth Ken ·· 021

Sustainable Forest Management and Rural Development in Lao PDR
    Airyai Vongxay ·· 028

Degraded Forest Rehabilitation Initiatives in Peninsular Malaysia
    Jennifer AnakFrancis ·· 035

Degraded Forest Rehabilitation and Sustainable Forest Management in Myanmar
    Chaw Chaw Sein ·· 047

Sustainable Forest Rehabilitation and Management for the Conservation of Trans-Boundary Ecological Security in Montane Mainland Southeast Asia: Pilot Demonstration Project in Myanmar
    Chaw Chaw Sein ·· 059

Potential and Prospects of Community Forest Management in Nepal: a Case Study of

Panchase Community Forest
  Rom Raj Lamichhane ··· 067

Degraded Forest Rehabilitation and Sustainable Forest Management in the Philippines
  Aurea P. Lachica ·· 074

Rehabilitation and Restoration of Forests in Sri Lanka
  W.T.B. Dissanayake ·· 082

Development of an Integrated Forest Management in Thailand
  Utharat Pupaiboon ··· 093

Degraded Forest Rehabilitation and Sustainable Forest Management: SFM Activities in Forest Industry Organization in Thailand
  Wondee Supprasert ··· 106

An Overview of Forest Rehabilitation in Vietnam
  Nguyen Tuong Van ·· 110

Restoration and Sustainable Management of Forest Ecosystem in the Central Highlands of Vietnam
  Nguyen Tuong Van ·· 124

Ecological Restoration of Degraded Forest Systems in the Tropics
  Shen Lixin, Jaap Kuper ·· 133

An Overview of Rehabilitation of Degraded Forest Lands in Asia

Dr. C.T.S. Nair[1], Prof. Shen Lixin[2] and Pan Yao[3]

Abstract

Deforestation and forest degradation remain major problems for most of the Asian economies and notwithstanding the ongoing efforts to counter them, the problems remain very critical. In the context of climate change mitigation and adaptation, rehabilitation of degraded forest lands has become a priority for most economies. The APFNet Workshop on Degraded Forest Rehabilitation and Sustainable Forest Management organized during 1 to 12 July 2014 in Kunming, Yunnan Province, China provided a unique opportunity to share forest rehabilitation experience from several economies in Asia. This paper provides an overview of the key findings from the papers presented during the Workshop, highlighting the nature and scale of the degradation problem, the causes of degradation, the impacts thereof, the ongoing efforts to rehabilitate degraded forest lands and the important lessons learnt.

Introduction

Degradation of forest ecosystems remains a major problem in almost all economies and this is particularly severe in the more densely populated tropics. The Global Partnership on Forest Landscape Restoration estimates that there are some 2 billion ha of land that needs restoration globally (https://www.iucn.org/about/work/programmes/forest/fp_our_work/fp_our_work_thematic/fp_our_work_flr/approach_to_forest_landscape_restoration/

---

1. Natural Resources Management Consultant, India
2. Director of APFNet Kunming Training Centre/Southwest Forestry University, China
3. Programme official, APFNet Kunming Training Centre/Southwest Forestry University, China

global_restoration_opportunity/) . As per an estimate of the International Tropical Timber Organization (ITTO, 2002), there are 500 million ha of degraded primary and secondary forests in the tropics. In addition, there are 350 million ha of tropical forest land, which is so degraded that forest regrowth has not occurred and which are mostly occupied by grasses and shrubs. In South East Asia alone, about 117 million ha or over 50 percent of the forest land is considered as degraded.

FAO defines forest degradation as "changes within a forest that affect the structure and functions of the stand or site and thereby lower its capacity to supply products and services" (FAO, 2011). The major concern therefore is the impact of degradation on human well-being through reducing the flow of goods and services. Vast tracts of forests are unable to produce products and services to their full potential. This is particularly a cause of concern considering the increasing demand for food, fuel, fibre and a whole range of ecological services like maintaining and improving watershed values, improving carbon sequestration and storage, protecting biological diversity and enhancing the aesthetic values of landscapes. The urgency of rehabilitating degraded forests has become particularly important in the context of climate change adaptation and mitigation to reduce carbon emission, a significant share of which is contributed by deforestation and forest degradation.

Efforts to rehabilitate degraded lands have a long history and a wealth of experience has been gained based on the work done during the last many decades. The *ITTO Guidelines on Restoration, Management and Rehabilitation of Degraded and Secondary Forests* (ITTO, 2002) outlines the principles and actions at two levels, namely (a) policy, planning and management level and (b) stand level. Most often the outcomes of restoration/rehabilitation efforts have been mixed and both successes and failures provide an opportunity to make refinements in policies and strategies. There is an urgent need to improve the quality of forest restoration/ rehabilitation at the site/landscape level and to find effective ways to undertake restoration in the broader environmental, social and economic context. Restoration of productivity is in fact an integral part of rebuilding the natural capital to ensure that the present and future generations are able to enjoy the full range of goods and services produced by the land.

It is in this context that the APFNet organized the Workshop on Degraded Forest Rehabilitation and Sustainable Forest Management in Kunming, Yunnan Province, China from 1 to 12 July 2014 with the following objectives.

- To assess the current state of rehabilitation of degraded forests in the Asia-Pacific region, particularly focusing on East Asia and South East Asia.
- To analyze the economic, social and environmental issues involved in forest degradation and their implications on rehabilitation/restoration efforts.

- To examine the future scenarios for forest rehabilitation taking into account the major drivers of ecosystem degradation, giving due attention to emerging policies relating to climate change mitigation and adaptation.

- To provide an opportunity to share knowledge on rehabilitation experience in the participating economies and to identify future options, particularly drawing upon successes and failures and the emerging approaches for adopting ecosystem approaches for restoration.

A key component of the Workshop was presentations by the participants on different aspects related to rehabilitation of degraded forest lands in the different economies. In all thirteen papers, including those outlining economy-wide status of degradation and rehabilitation, specific case studies were presented. This document includes edited versions of all the full papers presented by the participants.

Below is a synthesis of the key issues discussed in the papers largely focusing on the causes of degradation, ongoing efforts to rehabilitate the degraded forest lands and the key lessons – take home messages – as outlined in the different presentations.

The scale of degradation

Almost all papers presented during the Workshop outlined the severity of forest degradation, which although varies in degree, remains a major challenge to most economies. However there is a dearth of quantitative data on the scale and magnitude of the problem. Only two papers provided some indication of the scale of the problem—for example, in Lao PDR the extent of degraded natural forest that needs rehabilitation is estimated as 6.13 million ha, while in Peninsular Malaysia the extent of poor and degraded forests is estimated as about 0.30 million ha. None of the other papers provided any indication of the scale of degradation problem, notwithstanding the severity of the problem. Primarily this stems from the inability to define and measure degradation process, which is often a gradual and less perceptible process unlike deforestation. Also there are differences in the perception of what constitutes degradation. At times an area that may be considered as degraded by one group of stakeholders may be seen as a productive land by another group of resource users.

Drivers of degradation

Almost all the papers provided an indication of the causes of degradation, which in the ultimate analysis is linked to the collective impact of factors like population growth, economic changes and policy, legal and institutional factors. Population growth and increasing demand for products—especially woodfuel—is stated to be a key driver of degradation in Bangladesh. The situation in other low—income economies which are highly dependent on agriculture—for example, Cambodia, Lao PDR, Nepal, Myanmar and Vietnam—is the same. In the case of Cambodia, it is noted that some "75 percent of the rural population are farmers or land-less families who are highly dependent on access to natural resources for essential products, energy and food particularly in times of hardships". Increased demand with very limited resources to sustainably manage forests inevitably results in degradation, largely through multiple factors like over-exploitation, illegal logging, shifting cultivation, expansion of agricultural land and fuelwood collection. Urbanization is also identified as an important factor contributing to degradation. Although causes of deforestation and degradation are complex, the root cause is socio-economic (Box 1). Uncontrolled logging has been blamed as the principal cause of deforestation and degradation in many forest rich economies—for example, Philippines and Thailand—turning them from wood exporters to wood importers. War has been cited as an important cause of deforestation in the case of Vietnam; it is pointed out that during the two wars—1945 – 1954 and 1961 – 1975, Vietnam lost about 2 million ha of forests often due to the use of defoliating chemicals. Fire is cited as another key factor contributing to degradation as in the case of Malaysia, much of which is attributed to cultivation of oil palm.

> **Box 1**
> **Sri Lanka—causes of deforestation and degradation**
>
> "A complex web of factors, many of which are outside the forest sector, contribute to deforestation and forest degradation in Sri Lanka. Obviously the root cause is socio-economic in nature with poverty associated with landlessness and a poor land tenure system being the primary drivers. Other causes of deforestation are large agricultural and human settlement projects, such as Mahaweli Project, along with its reservoir and hydropower Projects, chena (shifting) cultivation, excessive harvesting of forest products, and the conversion of natural forests to plantation and arable land".
>
> Source: Dissanayake 2015.

Rehabilitation of degraded forests: it's evolution

Rehabilitation of degraded forests has a long history with some of the economies making rehabilitation efforts more than a century ago. Since then, the objective, approach and the policy, legal and institutional arrangements as also the state of technology have undergone changes as rehabilitation efforts attempted to adapt to the specific socio-economic conditions. Evolution of rehabilitation efforts in the selected economies are summarised below:

Early forest rehabilitation efforts

Almost all the Asian economies have a long and often chequered history of forest rehabilitation. Most often the history of reforestation/ restoration/ rehabilitation is linked to the history of forest management and economies with long established institutional framework in the form state of forest departments—for example, Malaysia, Philippines, Sri Lanka and Thailand—initiated rehabilitation efforts many decades ago. Some of the key features of early rehabilitation/ restoration efforts are described as follows.

- Rehabilitation initiatives have largely been designed to restock degraded areas particularly focusing on high-value species. Much of the thrust has been on high-value timber species like teak or veneer-log producing species in the tropical rain forests. The concern about a decline in the volume of commercially important species has been the prime driver of most forest rehabilitation efforts. It is in this context that techniques like assisted natural regeneration including line and gap planting in logged over areas—especially where selective logging has been undertaken—evolved.

- Most of the rehabilitation efforts were centred on government-managed lands especially those under the forest departments. Consequently, issues like tenure did not hinder the implementation of rehabilitation efforts.

- Of course, situations did exist where traditional land uses—especially shifting cultivation—led to conflicts. It is in this context that the taungya system or its variants were evolved and adopted widely in many economies. In addition to helping to resolve land use conflicts between shifting cultivation and industrial forestry, taungya also became a low-cost option for reforestation/ afforestation/ rehabilitation, helping to mobilise labour, especially in areas where labour supply was limited. It is for this reason that taungya which was first developed in Myanmar spread to several other economies. Although its success has been extremely varied, in a way it became a pre-cursor of participatory approaches to forest rehabilitation.

Expansion of industry-linked rehabilitation

With the decline in wood supply from natural forests and the increase in demand for industrial wood—especially for pulp and paper production—there has been a surge in rehabilitation efforts, largely focused on fast-growing species like eucalyptus, pines, acacias and such other species. Largely this has been a post-1960 phenomenon. The key characteristics of such rehabilitation/ restoration are as follows.

- The main thrust was on fast-growing short-rotation species, mainly intended for large-scale production of industrial raw material.

- Rehabilitation efforts began to be extended to areas outside forests, especially to degraded sites. With exotics gaining prominence, artificial regeneration became very critical and in a way natural regeneration of indigenous species faded out of favour.

- Substantial efforts were directed to develop nursery, planting and site management techniques with the primary objective of enhancing productivity. A wide array of technologies relating to production of planting materials, site management and management of pests and diseases were developed to address the multitude of challenges relating to rehabilitation of degraded forest lands, all focused on higher productivity.

Rehabilitation for meeting local community needs

A number of papers presented during the Workshop provided an insight into rehabilitation focused on meeting community needs, which obviously requires active involvement of the local communities. Several of these—in particular those discussing the situation in Lao PDR, Nepal, Philippines and Sri Lanka—underscore the importance of local community involvement in the rehabilitation of degraded forests. Meeting the local needs—especially woodfuel, fodder and small timber—remains the main thrust of most community-driven rehabilitation efforts. While meeting basic household needs has been the main focus of community-managed rehabilitation, this is changing and increasingly communities are also getting involved in the production of industrial wood as also in the provision of ecological services depending upon the specific context. The paper on Panchase community forest in Nepal provides an indication of the potentials and challenges in community management, especially when (a) resource management aims to accomplish multiple objectives, (b) human resource capacity is limited, and (c) management systems that help to resolve conflicts are yet to be developed.

Developing local capacity in managing natural resources thus becomes a key concern as elaborated in the Lao PDR paper. Considerable efforts are being made orient the staff and villagers and enable them to work in a co-operative framework, preparation of village development plans, management of funds including account keeping and so on. Such capacity development efforts are critical to ensure that local community involvement really succeeds.

Sri Lanka has also pursued the people's participation option, broadly referred to as social forestry approaches. Cooperative reforestation (which has been renamed as village reforestation recently) and farmer's woodlots are two important components of social forestry approaches being pursued in Sri Lanka to rehabilitate degraded lands under government ownership. The cooperative reforestation programme is based on the taungya system developed in Myanmar. The roles and responsibilities of farmers and forest departments have been well defined with the farmers getting the right to cultivate agricultural crops in government land for a period of up to 4 years in return for planting and maintaining teak plantation in the area. The farmer's woodlots programme has addressed some of the problems of taungya system, especially by way of increasing the lease period to 25 years and a variety of tree crops – including fruit trees are permitted to be grown.

Rehabilitation for enhancing the provision of environmental services

Although rehabilitation of degraded forests for provision of ecological services has a long history, it has gained momentum during the last two to three decades. An environment conscious society is demanding more green space especially to protect watersheds, conserve biodiversity, sequester carbon and enhance amenity values. Increasing potential of eco-tourism has generated considerable interest in rehabilitation focused on improving local livelihoods. This shift towards provision of environmental services has brought about fundamental changes in the concept of rehabilitation. Rehabilitation/restoration focused on environmental services has gained momentum in the context of addressing climate change. In fact, most of the national rehabilitation programmes—for example, the National Greening Programme (NGP) of the Philippines—has been launched in the context of climate change mitigation and adaptation strategies.

Most of the recent initiatives on rehabilitation of degraded forests aim to fulfil multiple objectives with considerable emphasis on provision of ecological services. Increased awareness about the ecological functions of forests have encouraged a shift away from the traditional industry-focused rehabilitation. In some cases, this has even led to the conversion of monoculture plantations into mixed stands by permitting the natural growth of local species.

From single objective to multiple objectives

In line with the principles of sustainable forest management, most rehabilitation programmes currently under implementation aims to accomplish economic, social and environmental objectives. The relative importance given to the different objectives differ especially in the context of who is actually managing the forest land. Establishing trade-offs between competing objectives remains extremely challenging and often this is addressed through grouping forests into different types. Thus while there is a broad consensus on the need to pay attention to social and environmental objectives, long term economic viability will require giving thrust to the production objective.

Policy, legal and institutional issues

Almost all the presentations underscored the importance of providing a robust policy, legal and institutional framework to promote forest rehabilitation. In fact, creating a favourable policy and institutional environment is the most important requirement—much more important than the technical aspects of rehabilitation. Most economies have revised their forest policies and legislation accordingly. Meeting basic needs of people and people's participation have become key thrust areas of many forest policies. Decentralized resource management especially through village-level institutions is being given emphasis and there are several national initiatives that aim to strengthen local capacity to manage natural resources sustainably. New forest policies also encourage private sectors' involvement in afforestation and reforestation. Some of the major initiatives—for example, China's Sloping Land Conversion Programme and Vietnam's Five Million Hectare Restoration Project (Programme 661) and Greening the Barren Hill Program (Program 327)— involved major policy, legal and institutional changes.

> **Box 2**
> **Basic principles of Cambodian National Forest Programme**
>
> The Cambodian National Forest Programme is to serve as an appropriate mechanism to provide a transparent and participatory process for planning, implementation, monitoring, evaluation and coordination of all forestry activities adhering to the following principles.
>
> - Sustainable forest development adhering to social, economic, cultural and environmental aspects;
>
> - Effective economy leadership and good forest governance: conflict management, commitment and alignment with national policies;
>
> - Regular participation through multi-stakeholder consultation: technical working groups, technical assistance and partnerships appropriate to the Cambodian context;
>
> - Holistic and cross-sectoral approaches: using landscape planning approach through collaboration with relevant government agencies, local government and civil society;
>
> - Monitoring mechanism: its implementation for improved performance and for public information and awareness raising among stakeholders including national and local governments, civil societies and the private sectors.

Many of the policy, legal and institutional changes have been outlined in the strategies and action plans, including the national forest programmes. And often the national forest programmes have strong linkages with the overall national development plans. For example, in the case of Cambodia, the national forest programme is guided by the millennium development goals of poverty reduction, environmental sustainability and attainment of a forest coverage of 50 percent of the land area by 2015 (Box 2).

Tenure reform: a critical requirement for successful forest rehabilitation

The various studies—including case studies—indicated the critical role of tenure reform, especially to encourage local community's involvement. In almost all economies which have made significant progress in forest rehabilitation, tenure reform has been a key step as in the case of China, Nepal and Vietnam. Most economies in Asia have adopted a variety of measures to confer tenure security so that private land owners and communities will have the necessary incentives to invest in rehabilitation and sustainable forest management (Box 3). Tenure reforms include short term to long term leases of public land as also various forms of public-private partnerships.

> **Box 3**
> **Community forestry in Myanmar**
>
> The Myanmar Forest Department has issued the *Community Forestry Instructions* (CFI) in 1995 so as to develop community participation in forest management in Myanmar. CFI allows local communities to be involved in developing management plans, protection, conservation and restoration of forests, and encourages the utilization of forest products and Non-Timber Forest Product particularly in the vicinities of their settlements. The CFI allows for a 30-year land use, ownership rights and disposal of products from community forest under the guidance of Forest Department.

Funding forest rehabilitation

Although financing rehabilitation remains a challenging issue, only a few papers presented during the Workshop (for example Vietnam—see Box 4) addressed the issue. Preventing degradation requires that removal of products is limited to what is sustainable and the structure and functions of forests are maintained in tact. In the context of increasing human pressure, this becomes a challenging task. As such most of the rehabilitation is funded by governments either directly or indirectly. Several economies have also used significant external funding from bilateral and multilateral organizations to undertake rehabilitation.

Involvement of other stakeholders, especially local communities, farmers and private investors, is often intended to reduce the burden on public funding. However, in many cases local communities require significant public funding to pursue rehabilitation efforts, especially in situations where communities are poor or where rehabilitation is primarily focused on enhancing environmental services. In this regard, it has to be noted that most of the so-called PES schemes—which aim to generate income through provision of ecological services like watershed protection, carbon sequestration, *etc.* —operate based on direct and indirect public subsidies.

> **Box 4**
>
> **Funding forest rehabilitation—the situation in Vietnam**
>
> "Vietnam has for many years invested a significant amount in forest rehabilitation, especially since the 1990s. This national investment has been complemented with substantial international support.
>
> Under the current arrangements of payments for the protection of forests, state financing of forest protection needs to continue to ensure the effective accomplishment of the objectives. There is little other funding for forest rehabilitation, especially for the rehabilitation of production forest land that is meant to boost the forestry sector's contribution to the national economy.
>
> This funding situation does not translate to optimal conditions for smallholders. Some payments, such as for forest protection contracts, are perceived to be too low. Credit available for forest rehabilitation needs to be more favourable; but even when credit availability is not a problem, convincing farmers to invest in forest rehabilitation remains challenging."

The case studies also indicate that unless there is strong political commitment, public funding of degraded forest rehabilitation becomes extremely challenging considering other governmental priorities.

Obviously another option is to encourage private investments in forest rehabilitation. This however requires changes in policies and legislation and rehabilitation should be an economically viable option. Probably the industrial model of rehabilitation focused on wood production is more likely to attract private funding. Rehabilitation for fulfilling social and environmental objectives will largely remain in the realm of public funding and given the limited ability of economies to mobilise resources, this will continue to be quite difficult.

Creating public awareness and social mobilisation

Almost all papers provide an indication of the ongoing efforts to generate public awareness about the causes and consequences of forest degradation and what needs to be done to reverse the process. In many economies, tree planting has gained prominence on account of celebrations associated with Arbor Day. For example, Malaysia has launched several tree-planting campaigns—26 Million Tree-Planting Campaign, the International Day of Forests, International Year of Forests 2011, *etc.* —involving government agencies, private sectors, civil society organizations, schools and universities and public at large to underpin the need for ecosystem rehabilitation. Some of these efforts have gained considerable momentum especially when there is strong political commitment backed up by resources and institutional capacity as in the case of the National Greening Programme in the Philippines.

Key messages

Overall the papers provide a glimpse of the diverse aspects of degraded forest rehabilitation indicating how they have evolved, changes in rehabilitation objectives, the roles and responsibilities of different players/stakeholders, technological developments and so on. Most economies have very rich experience in taking up rehabilitation efforts. Some of the key conclusions stemming from the different papers include the following.

- Most of the degradation is taking place in public lands. This largely reflects a situation where most forests are under public ownership and that in most cases governments face complex problems in managing these forests sustainably. Removal of products often far exceeds what is actually sustainable. A major challenge facing public forestry institutions—especially forestry departments—is the inability to exclude anthropic pressures, in particular in densely populated areas. Unsustainable and illegal removal of products, encroachment, grazing, fires, *etc.* are particularly severe in such areas accentuating the degradation process. And obviously public investment in rehabilitation/sustainable management has been far lower than what is actually removed.

- Most economies in Asia (as elsewhere) are fully aware of the consequences of forest degradation and have initiated measures to restore/rehabilitate the degraded lands. One widespread response has been to impose a ban on logging—as in the case of Bangladesh, China, Philippines, Sri Lanka and Thailand. Whether such bans have actually helped in arresting degradation and deforestation is a moot question.

- The case studies provide a clear indication of how rehabilitation efforts have undergone changes, especially as regards the objectives and the stakeholders involved. In most cases rehabilitation initiatives in the early years have largely focused on industrial wood production. However, this has changed significantly as the stakeholders involved are much broader and there is a growing concern about the decline in environmental services and social services provided by forests.

- There is a clear recognition of the importance of participatory approaches in the success of rehabilitation efforts. This is especially so in economies where population pressure is high and no meaningful rehabilitation can be accomplished through excluding people. Most of the successful stories of rehabilitation are from those economies that have put in place effective participatory approaches.

- Increased participation often requires fundamental changes in policies, legislation and institutions. As already pointed out, most rehabilitation efforts earlier have been top-down initiatives appropriate to industrial model of wood production. A shift in focus towards multi-purpose rehabilitation will require changes in policies, legislation and institutions.

- Funding rehabilitation will continue to be a major challenge for most economies. Hitherto most rehabilitation has largely relied on public funding, both national and international. Funding from the private sector is primarily dependent on financial viability of rehabilitation and is more likely to be focused on industrial type forestry. Although some efforts have been made to develop systems to pay for environmental services (PES) to support degraded forest rehabilitation, their overall impact has not been encouraging and in many cases PES is driven by direct and indirect support from governments.

- Often the scale of the problem is so large, social mobilisation involving all stakeholders becomes critical. Some of the successful stories—for example, Korea, China and Vietnam—involved significant social mobilisation through a combination of efforts going far beyond the usual "carrots and sticks" approaches and making nature conservation a mass movement.

References

FAO. 2011. Assessing forest degradation: towards the development of globally applicable guidelines. In: FAO. Forest Resource Assessment Working Paper, 177. Rome: Food and Agriculture Organization of the United Nations.

ITTO. 2002. ITTO guidelines for restoration, management and rehabilitation of degraded secondary forests. In: ITTO. ITTO Policy Development Series. International Tropical Timber Organization, http://www.itto.int/policy papers_guidelines/.

2

Participatory Forest Management in Degraded Forests: Perspective REDD+ in Bangladesh

Prof. AL-Amin, M[1] and Prof. Hasan, M[2]

Abstract

Although British colonial approach still forms the foundation of forest resource management in Bangladesh, there are several signs of change. Forest management policies are increasingly incorporating social aspects giving due consideration to diverse human dimensions. Concern about climate change and the need to improve performance of afforestation/reforestation programs, REDD+ *etc.*, have encouraged a significant shift from policies pursued earlier. Based on a case study in the Teknaf forest region (Southeast Region), this paper critically reviews the forest management practices and the potential for participatory management approaches in the context of implementation of REDD+. The results reveal that people in the study area are conscious about the forest products and services and are themselves getting involved in law implementation including guarding the forest resources from illicit felling. This case study provided an indication of how a participatory management approach might help in the rehabilitation of deforested and degraded areas indicating a better option in the implementation of REDD+ in Bangladesh.

Introduction

Land use in Bangladesh is dominated by agriculture (64.2 percent), and forestry (17.8

---

1. Corresponding author, Institute of Forestry and Environmental Sciences, University of Chittagong, Chittagong 4331, Bangladesh. E-mail: prof.alamin@yahoo.com.

2. Institute of Forestry and Environmental Sciences, University of Chittagong, Chittagong, Bangladesh

percent) (Figure 1). Homestead forests and government-owned forests are the key sources of fuel and timber. However, rapid population growth and increased demand for fuelwood and timber coupled with shifting cultivation, illicit felling and forest fires have accelerated soil erosion and consequent degradation of land. Further, the economy is regularly affected by natural calamities like floods, cyclones, *etc*. Encroachment of forest lands is also a major problem. Addressing these have necessitated a paradigm shift in forest management—from the top-down approach developed during the colonial times to the adoption of participatory approaches, especially considering the high degree of dependence of local people on forests. The government of Bangladesh has imposed a ban on cutting trees since 1989. Although home gardens and imports meet a significant share of wood demand, illicit felling from government forests remains a major problem accentuating forest degradation. This situation requires a detailed assessment of how these areas could be rehabilitated, and how programmes like REDD+ could counter the degradation challenge.

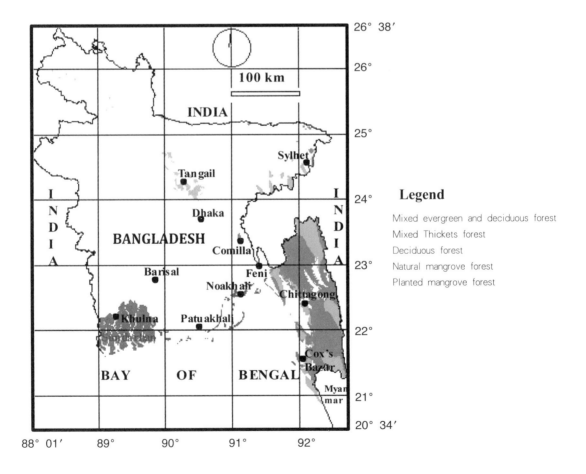

Figure 1 Forest cover in Bangladesh (extracted from Al-Amin, 2011)

Methodology

Description of the study area

Cox's Bazar South Forest Division is situated in the extreme south eastern region of Bangladesh. It lies between 20°50' and 21°51' N latitude and 92°0' and 92°15' E longitude. Teknaf is an upazilla of Cox's Bazar, the southeast district in Bangladesh, bordering Myanmar. The area of forests and the staff strength in the Cox's Bazar South Forest Division are given in Table 1.

Table 1 Forest area and staff strength in Cox's Bazar South Forest Division

Range	Reserved Forest (ha)	Protected Forest (ha)	Total area (ha)	Number of beats	Number of staff
Teknaf	6011.13	613.20	6624.33	4	15
Hoaikong	5186.29	10.87	5197.16	4	14
Silkhali	2956.27	22.21	2978.48	3	12
Total	14153.69	646.28	14799.97	11	41

Source: Teknaf Range Office, Cox's Bazar South Forest Division.

Data collection

Resource survey

Resource survey was conducted in three beat areas of Teknaf forest region, namely Teknaf, Hoaikong and Silkhali ranges. Simple random sampling was followed to collect the resource data adopting a predetermined sample size. Sample plots with 20m × 20m were taken in the forest for collecting diameter at breast height (dbh) and height. In the case of large plantations, the sample plot size was 50m × 50m. For regeneration survey, a smaller plot size of 5m × 5m was adopted. Tree volume was calculated and the collected data were analyzed to prepare local volume tables for the different species.

Socio-economic survey

A household survey was conducted in different villages of Teknaf, Shilkhali and Hoaikong forest range of Cox's Bazar South Forest Division to assess people's dependency on forests and the economic status of households. The survey was conducted during April 2014 using a structured questionnaire. In each household, the head of the household (or alternatively the female head or any adult member of the family) was interviewed. The collected data were analyzed and presented in the result section.

Results

Natural forests

Cox's Bazar South Forest Division holds a very small area of productive natural forests. It is situated at Shilkhali Range. *Dipterocarpus alatus* (Dhuli Garjan) (5,717 trees) is the dominant species and *Hopea odorata* (Telsur) as the co-dominant species. The forest is about 150 years old. The average growing stock in the forest is 340.9 m^3/ha (Table 2). Regeneration of these forests is satisfactory.

Table 2   Dhuli Garjan (Dipterocarpus alatus) Forest at Silkhali

Range	Location	Dominant species	Volume (m^3/ha)
Silkhali	Silkhali	Dhuli Garjan: *Dipterocarpus alatus*	340.9

Plantation forests

Most of the study area consists of uneven-aged multi-storey plantation forests, with a canopy height of 25 to 35 metres. Evergreen species are frequent in the upper storey. Important species in the main canopy are *Dipterocarpus turbinatus*, *Dipterocarpus alatus Termanila arjuna*, *Chukrassia tabularis*, *Syzygium grande*, *Tectona grandis*, *Gmelina arborea*, *Acacia auriculiformis* and *Casuarina equisatifolia*.

Among the plantation forests, *Acacia auriculiformis* (Akashmoni) plantations (Figure 2) are found in Kerontoli, Domdomia and Nature Park of Teknaf Range which hold 93.9 m^3/ha, 81.31 m^3/ha and 214.64 m^3/ha of timber respectively. Local people involved in the co-management of the plantations prefer this species in view of its gregariousness, low maintenance cost and low vulnerability to grazing. However, being an invasive species, *Acacia auriculiformis* supresses the growth of indigenous tree species. Other plantations in the area are Agar plantation at Teknaf Sadar, Teak plantation at Kerontoli, Sal plantation at Boroitoli and Jhau plantation at Silkhali which hold 89.09 m^3/ha, 7.1 m^3/ha, 30 m^3/ha and 104.9 m^3/ha of timber respectively (Table 3).

Figure 2   *Acacia auriculiformis* **plantations**

Plantations of Jhau (*Casuarina equisetifolia*) are in the coastal areas of Shaplapur and Silkhali. Stocking is poor in some places due to grazing and other biotic interferences. The Shaplapur Jhau plantation has a mean dbh of 15.7 cm. Number of trees average 1,903 trees/ha and standing volume is 222.9 m^3/ha. Most of the trees are in the 18 – 19 cm

Table 3   Volume of plantation forest of teknaf wildlife sanctuary

Location	Planted species	Volume (m³/ha)
Teknaf Sadar	Agar (*Acquillaria agallocha*)	89.09
Kerontoli	Akashmoni (*Acacia auriculiformis*)	93.90
Kerontoli	Teak (*Tectona grandis*)	7.10
Domdomia	Akashmoni (*Acacia auriculiformis*)	81.31
Boroitoli	Sal (*Shorea robusta*)	30.00
Silkhali	Jhau (*Casuarina equisetifolia*)	104.90
Nature Park	Akashmoni (*Acacia auriculiformis*)	214.64
Shaplapur	Jhau (*Casuarina equisetifolia*)	229.90
Silkhali	Jhau (*Casuarina equisetifolia*)	209.30

diameter class, while the maximum height and diameter of the trees are 18.9 m and 21.1 cm respectively. The Silkhali Jhau plantation, has a density of 1,763 trees/ha with a mean dbh of 14.2 cm. Number of trees average from 408 trees/ha and standing volume is 209.3m³/ha. Maximum numbers of trees are of 16 – 17 cm diameter class. The highest and lowest heights and diameters of the trees are 23 m and 3.5 m, 26 cm and 4.7 cm respectively.

Bamboo resources

Bamboo plantation in Teknaf Sadar Beat under Teknaf Range has an average of 791 culms/ha. In addition, several other species of bamboo are also found in these forests. The most common bamboo species are: Muli (*Melocanna baccifera*), Mitinga (*Bambusa tulda*), Dolu (*Neohouzeaua dulloa*), Ora (*Dendrocalamus longispathus*), *etc*.

Cane resources

Cane has been planted on an experimental basis in some areas of Teknaf and Hoiakong range. A total 110 plantations have been established in Hoiakong range and in Teknaf Sadar Beat. Cane is found as an associate species along with *Chuckrassia tabularis* (Chickrashi) at Teknaf Sadar Beat under Teknaf Range which holds 381 clumps/ha. Cane species occur in association with Chickrassi, Teak, and Koroi, *etc.* (Figure 3, Figure 4). The common cane species found are as follows: Golla bet (*Daemonorops jenkinsianus*), Kadam bet (*Calamus errectus*), Sundi bet (*Calamus guruba*), Udum bet (*Calamus longisetus*), Kora bet (*Calamus latifolius*).

Socio-economic status

Educational status of the respondents shows a high level of illiteracy with 60 percent being illiterate. Among the literate, 27.5 percent have primary education and 8.75 percent have

Figure 3 *Termanila arjuna* with cane

Figure 4 *Chukrassia tabularis* with cane

secondary education. Family status of the respondents shows that average family size in the study area is 6.25 of which 2.5 are males and the remaining females. On an average, each family has 2.25 earning members all of whom are males. Analysis of the family income by the respondents households show that average family income in the study area is 13,000 Tk/month (US$167) and 52,000 – 55,000 Tk/year (US$667 – 705) of which 47.5 percent is obtained from fishing. Tables 4 and 5 summarize some of the results of socio-economic survey conducted in the study area.

Table 4   Educational status   %

Location/village name	Illiterate	Literate					
		Pri	Sec	S.S.C	H.S.C	Gra	Total
Silkhali	40.00	45.00	15.00	0.00	0.00	0.00	60.00
Teknaf	30.00	40.00	15.00	10.00	5.00	0.00	70.00
Shaplapur	80.00	15.00	5.00	0.00	0.00	0.00	20.00
Jailer dip	90.00	10.00	0.00	0.00	0.00	0.00	10.00
Average	60.00	27.50	8.75	2.50	1.25	0.00	40.00

Table 5   Dependency on forests   household

Area	Fuelwood	Timber extraction	Fuelwood + timber extraction	Seedling cutting
Silkhali	20			
Shaplapur	32	1	47	
Holbunia		20		
Dochakmapara	29			
Jahajpura		5		15
Madargunia		17		
Shilcharipara	23		1	
Total	104	43	48	15

Discussions

The natural forests which once dominated the study area are now under threat of intense exploitation. Some of the areas have been converted into plantations and currently the extent of natural forest stands is limited to about 750 ha with just over 5,700 trees giving an indication of the earlier forest composition. They exist on account of the protection given by the government forest department. If these areas are also under participatory system of forest management, the potential for carbon sequestration under REDD+ will be much higher.

Intensive management involving local communities could however significantly increase carbon sequestration as indicated in Table 3. As per the benefit sharing system, local communities get 45 percent of the income from selling the products from thinning and final felling (the remaining 45 percent goes to the forest department and 10 percent to cover the cost of reforestation and maintenance). All the plantations under participatory management yield a higher volume of wood than conventional management. Protecting the forests from illegal felling remains a major problem in conventional forest management (Figure 5).

Since 1989, the Government of Bangladesh has imposed a ban on logging forests under government ownership; however trees grown under participatory management can be cut and this has encouraged the local communities in protecting trees under participatory management. In the Teknaf area, 505 ha of plantations under participatory management is proposed to be felled and this could generate substantial income to local communities (Figure 6). Local communities are actively involved in the management and protection of these areas.

Figure 5 Illicitly felled area under conventional management practice area for natural forest

The socio-economic study has revealed that the people of the study area are not highly educated and illiteracy rate is very high which increases the dependency of the people on the adjacent forest for their livelihood. Participatory forest management taking advantage of REDD+ opportunities will not only conserve the forest as a carbon sink but also help in improving the socio-economic conditions of the local communities (Figure 7).

Figure 6   Beneficiary with dao

Figure 7   A patch of natural forest need to be conserved

Conclusions

Bangladesh is a densely populated economy where land is one of the scarce resources. As forests are indispensable for livelihood, they are being depleted. This case study clearly indicates how participatory management may help forest reconstruct from its deforested and degraded status, and hence, act as a better option for the implementation of REDD+ in Bangladesh.

Acknowledgements

The authors wish to acknowledge the support provided by the students of the 8th semester (2008 – 2009 session), Prof. G.U. Ahmed and Mr. T.K. Bal, an assistant professor of IFESCU for undertaking field work and data collection that enabled the preparation of this paper.

References

Al-Amin, M. 2011. Application of spatial data in forest ecology and management. In: Lambert Academic Publications. Germany, 242 pp.

Forest Restoration and Plantations in Cambodia

Piseth Ken[1]

Abstract

The paper provided an overview of the present state of forests and forestry and the ongoing efforts to implement sustainable forest management under the National Forest Programme 2010 – 2029. Cambodia had witnessed continuous deforestation, especially on account of unregulated logging. However, there were signs of improvement especially on account of governance improvement and the concerted efforts to implement the National Forest Programme. The National Forest Programme 2010 – 2029 had 6 integrated programmes addressing different aspects of sustainable forest management. Substantial effort had been made towards establishment of plantations involving different stakeholders. Creation of awareness among different sections in society was another priority area.

Background

Forest resources have been providing immense benefits to Cambodian society. In addition to direct economic benefits, it provides several services including climate regulation, biodiversity protection, regulation of water flow, protection of slopes, *etc*. Further, it also supports the livelihood of millions of people who rely on a number of products, in particular non-wood forest products.

---

1. Technical Officer of Department of Forest Plantation and Private Forest Development, Forestry Administration

Cambodia has succeeded in maintaining an extensive but varied forest extending over an area of about 10.8 million hectares, including bamboo forests and planted forests. These forests have unique biological features and they also form an important cultural heritage. Cambodian forests have also close linkages with agriculture and inland fisheries and help maintain the productivity of these two sectors.

The majority of the Cambodian population is rural and consists primarily of subsistence farmers, 75 percent of whom are marginal or landless families highly dependent on access to natural resource for essential products, energy and food, especially during times of hardships. Forests also provide household opportunities for diversification, supplementary income, and employment created by forest products-based enterprises.

Deforestation in cambodia

Earlier governments have used forests as a means of generating income to support their activities. As per an assessment in 1965, forests accounted for 73.04 percent of the land area. The system of giving large scale concessions in the 1990s led to illegal and unsustainable logging, reducing the forest area to 59.82 percent in 1993, 58.60 percent in 1997 and 57.07 percent in 2010 (Figure 1).

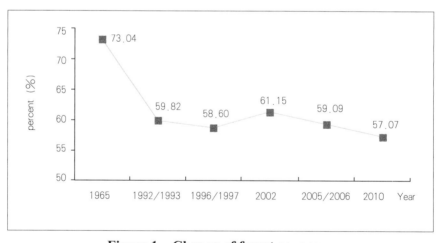

Figure 1 Change of forest coverage

National forest programme 2010 – 2029

The Royal Government of Cambodia has been implementing significant reforms in the forest sector since 1998 and has achieved remarkable results in building a good foundation for implementing sustainable forest management, contributing to social-economic development, environmental protection, and poverty alleviation, safeguarding the resources for future generations. In order to sustain effective forest reform, the Royal Government of Cambodia has embarked on an intensive process of developing and implementing the National Forest Programme 2010 – 2029 (NFP) aimed to place good governance and effective partnerships as the core of sustainable forest management.

The NFP will be guided, among others, by the Cambodian Millennium Development Goals for poverty reduction, environmental sustainability and attainment of 60 percent forest coverage by 2015. The NFP is designed to serve as an appropriate mechanism to provide a transparent and participatory process for planning, implementation, monitoring, evaluation and coordination of all forestry activities, adhering to the following principles.

- Sustainable forest development adhering to social, economic, cultural and environmental aspects.
- Effective economy leadership and good forest governance: conflict management, commitment and alignment with national policies.
- Regular participation through multi-stakeholder consultation: technical working groups, technical assistance and partnerships appropriate to the Cambodian context.
- Holistic and cross-sectoral approaches: using landscape planning approach through collaboration with relevant government agencies, local government and civil society.
- Monitoring mechanism: implementation to be monitored regularly for improved performance and for public information and awareness raising among stakeholders including national and local governments, civil societies and the private sectors.

The National Forest Programme 2010 – 2029 consists of six components each comprising a key aspect of Sustainable Forest Management (SFM) to be accomplished during the period from 2010 to 2029 as indicated in Figure 2.

Conservation and development of forest resource and biodiversity

Forests are an integral part of Cambodian life and culture. Most rural people are dependent on forest products for their livelihood. The Royal Government of Cambodia considers that the implementation of environmentally, socially and financially sustainable management of forest resources could contribute significantly to poverty reduction and socio-economic development. Currently, it encounters challenges in forest management due to factors such as unclear/ambiguous land rights, lack of demarcation and absence of proper management plans resulting in deforestation and habitat degradation. To address these deficiencies, the Cambodian government has put in place policies, strategies and tools to implement SFM to achieve the goal of 60 percent forest coverage by 2015 taking cognizance of the Cambodian

Programme I: Forest Demarcation, Classification and Registration	• Forest Demarcation, Forest Classification and Registation • National Function-based Forest Classificaiton
Programme II: Conservation and Development of Forest Resource and Biodiversity	• Forest Management Plan • Development and Management of Production Forest • Monitoring, Assessment and Reporting for SFM • Biodiversity and Wildlife Conservation • Conservation and Development of Genetic Resource and Seed Source • Three Plantation and Development of Forest Plantation • Development of Forest Products and Market Promotion • Wood Technology Development and Forest Product Processing • Forest Certification
Programme III: Forest Law Enforcement and Governance	• Legal and Adminstrative Reform • Law Enforcement and Forest Crime Monitoring and Reporting • Rapid Respone on Forest Crime Information • Conflict Management System • Monitoring, Reporting and Learning System
Programme IV: Community Forestry Programme	• Community Forest Identification and Formalisation • Community, Institutional and Livelihoods Development • Community Forestry Development Support
Programme V: Capacity and Research Development	• Institutional and Human Resource Development • Extension and Public Awareness • Research Capacity Building Development
Programme VI: Sustainable Forest Financing	• Government Financing • Income from Forest Sector • Income from the Private Sector and Community Foresty • Financing via Donors • Innovation Financing from Payments of Environment Services and Carbon Credit

Figure 2　Framework the National Forest Programme 2010 – 2029

Millennium Development Goals. Thus, the NFP has been formulated by the Royal Government of Cambodia to accomplish the following:

- To improve national land-use planning and national forest resource;

- To support implementation of forest management systems;
- To conserve genetic diversity and biodiversity;
- To enhance the benefits from forest services; and
- To support wood processing technology development, and enhance quality of forest products and market promotion.

Establishment of forest plantations is an important approach to arrest deforestation and habitat degradation and to increase the forest cover in addition to contributing to socio-economic development and poverty alleviation.

Development of forest plantations

The National Development Programme 2010 – 2029 in Cambodia recognizes the importance of forest resource to rural livelihood. The Forestry Administration's efforts to create an enabling environment to develop plantation forestry will focus on multi-purpose tree plantations, which have a potential to supply domestic timber needs, increase income to local communities and improve the environment through watershed protection and erosion control. It is important that these plantations are developed wisely giving consideration to multiple uses, grown on long and short rotations (for example, *Dipterocarpus alatus*, *Hopea odorata*, *Azadirachta*, rubber, *Eucalyptus* spp., and *Sesbania grandiflora*). These plantations are being developed taking into account the present and future market demand. Awareness of the long term benefits will help deal with the negative perception that some villagers may have against growing trees.

Forest plantation statistics of Cambodia

Forestry Administration, especially the department of forest plantation and private forest development, has been pursuing forest restoration and plantation development efforts in collaboration with relevant stakeholders such as government departments, civil society organizations and private sector in order to achieve the 60% forest coverage target by 2015. Since 1985 till 2011, Forest Administration (FA) has been able to raise 18,076 ha of plantations (Figure 3). The Department has 16 forest restoration and plantation stations, 25 nurseries and 6 agro-forestry pilot plots. These stations also encourage and support plantations of indigenous species and hitherto restoration involving species like *Cassia siamea*, *Dipterocarpus alatus*, *Eucalyptus*, *Hopea odorata*, *Peltophorum*, *Tectona grandis* and *Dalbergia cochinchinensis* have covered an area of 12,945 ha.

Arbor Day

Annually the Forest Administration celebrates the Arbor Day on 9 July and His Majesty the King Norodom Sihamoni plants young trees and offers saplings to a local resident to

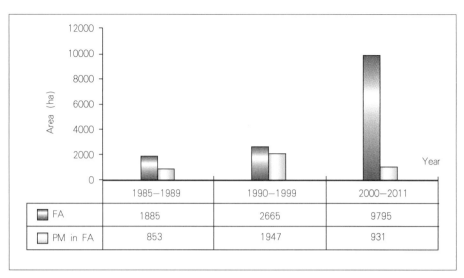

Figure 3 Forest Plantation Statistics

disseminate the message about the importance of forests. This event is attended by senior officials and representatives of relevant non-governmental organizations as well as thousands of local residents and students. In addition to the celebrations at the national level, the Arbor Day is also celebrated at various levels to convey the importance of forests and trees in the well-being of people and the economy as a whole.

Nurseries

Every year nurseries managed by the Forestry Administration distribute tree seedlings to the public to plant along village roads, pagodas, schools, *etc.* (Figure 4) The distribution is often done by senior officials including the Prime Minister (PM).

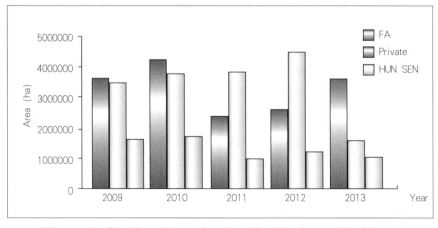

Figure 4 Seeding plantation for distribution to Residents

Challenges

Although Cambodia is making every effort to reduce deforestation and degradation and to increase the area under forest cover, many challenges persist. Land encroachment continues to remain an important problem affecting many of the forest restoration and plantation stations. There is also a need to improve cooperation and coordination between different organizations, especially non-governmental organizations (NGOs). Forest fires continue to be another persistent problem affecting the outcomes of restoration efforts.

Conclusions

The Forestry Administration hopes that forest development strategy will create an enabling environment for multipurpose tree plantations, which will increase the supply of timber, increase the income of local community and improve the environment through watershed protection and erosion control and eventually increase the forest cover to 60 percent as stipulated in the Cambodian Millennium Development Goals.

The National Forest Program 2010 – 2029 is expected to accelerate forest restoration and plantation efforts contributing to the Government's Rectangular Strategy Phase III aimed to improve agricultural productivity and diversification through land reform and fishery and forestry reforms. In particular attention is being focused on:

- Creation of an enabling environment for local investments in multi-purpose tree plantations through ongoing reforms to secure local people's rights to benefit from all aspects of plantation activities;

- Rural people's involvement in multi-purpose plantation increase steadily, and as a result improvement of local incomes and livelihoods in participating communities;

- Watershed protection and reduced soil erosion in participating area;

- Sustainable supply of timber and poles through production of wood products at local and national levels.

4 Sustainable Forest Management and Rural Development in Lao PDR

■ Airyai Vongxay[1]

Abstract

The paper provided an overview of forest conservation and rehabilitation efforts in Lao PDR, especially of arresting the pace of deforestation and degradation. Considering the high level of poverty and the dependence of local people on forests, rehabilitation and livelihood improvement efforts were fully integrated involving all the key stakeholders. An important component of this was to build the technical and managerial capacity of villagers, especially to plan, organize and monitor different activities.

Background

Lao PDR is a land-locked economy surrounded by China, Myanmar, Vietnam, Thailand and Cambodia. The total land area is 23.68 million ha out of which 16.5 million ha is classified as forest land. However, the area with forest cover is only 9.8 million ha or 41 percent of the land area. The total population is around 6 million, mostly (about 60 percent) living in rural areas. The forest is divided into three types: conservation, protection and production forests. The extent of area classified as production forest is about 3.1 million ha, while the extent of protection forests is 8.2 million ha. The mandate to manage the forests vests with the MAF/DOF (Ministry of Agriculture and Forestry/ Department of Forestry).

---

1. Technical Staff of REDD+ Office, Department of Forestry, Ministry of Agriculture and Forestry

Currently sustainable forest management is being implemented in conjunction with rural development by focusing on the rehabilitation of all three forest types mentioned earlier.

Management of production forests

Several steps are involved in the management of production forests which include (a) delineation of the management areas, (b) preparation and approval of forest management plan by government, (c) inventory of the area which will be divided into forest management areas and compartments, (d) data collection and analysis determining annual logging areas taking into account rotations, (e) pre-logging survey, (f) tree marking, (g) monitoring and evaluation by the Sustainable Forest Management Committee during logging operations, (h) rehabilitation planning, including natural reforestation in production forest areas, enrichment planting on degraded forest areas and planting on opened forest areas, (i) nursery activities including preparation of soil, sand, water, bags, seeds, fertilizer, tools and mobilising labourers, and (j) forest protection and seedling maintenance: weeding, replanting, and/or watering.

Degraded forest land rehabilitation

Vast tracts of natural forests have been degraded on account of several factors and systematic efforts are underway to rehabilitate them. The total area of the natural degraded forest that needs rehabilitation is estimated as 6.13 million hectares. For the period 2006 – 2010, the Forestry Department was able to rehabilitate about 614,000 ha out of a targeted area of 2.550 million ha.

Plantation and seed sources

There were about 100 seed production areas in the whole economy under the management of government having about 6,600 trees consisting of 28 species spread over an area of 9.188 ha.

Main problems of forest rehabilitation

There are several problems affecting forest rehabilitation efforts, the most important of which are:

- Inadequate understanding of the issues relating to rehabilitation at the provincial and district levels;
- Limited resources available for rehabilitation, which in particular is resulting in poor maintenance of areas planted;

- unclearly-identified and unclearly-delineated land for plantation investment by industries is not clearly identified and delineated ;
- Complexity of the process of approval for investments in plantations in view of the multiplicity of agencies as also complex approval process involving officials at various levels; and
- Also delay in approving the plan and the use of the Forest Development Fund for forest rehabilitation and tree planting.

Participatory village development activities

Lao PDR has mainstreamed participatory approaches in the rehabilitation of degraded areas with different components addressing different aspects as indicated below.

Capacity building of village organizations and staff

Ensuring that the village organization works effectively and efficiently in close collaboration with the district and project authorities is a key aspect of the success of rehabilitation and sustainable management of forests. The main thrusts under this are the development of human resources and skills to undertake the different tasks and strengthen the facilities and physical infrastructure required by the village organization to undertake the different tasks.

Land use zoning and planning

Land use zoning and planning aims to optimize the resource use to ensure sustainability and productivity, including the identification of forests to be managed by villagers and to systematically reduce the extent of area under slash and burn cultivation, eventually resulting in the complete cessation of such cultivation. Several steps are involved in the process, namely:

- Preparation and orientation of staff and villagers;
- Establishment of the Village and Group and Village Development Committee for land use planning;
- Delineation and verification of village boundary;
- Assessment of indigenous land management practices and historical change in land use;
- Land use and forest cover mapping by villagers themselves. which is facilitated by the use of satellite imagery interpretation followed by detailed participatory mapping and field checking by villagers.

- Socio-economic analysis and estimation of future land requirements for cropping, livestock, houses, infrastructure and forests for the whole village population for the next 20 years;
- Forest and land zoning, which is focused on conservation of medium and good quality forests (for Public Forest Areas or village forest), while identifying areas with potential for current and future uses for the livelihood needs of villagers;
- Preparation of village forest and agricultural land management agreements; and
- Demarcation of boundaries to indicate the different zones.

Village development planning

In addition to land and forest zoning, efforts are also directed towards supporting the overall socio-economic development of the villages, particularly focusing on poverty alleviation. The main objectives of this are (a) to guide the implementation of development activities which are effective in meeting the local needs, (b) to improve and develop the capacity of the village organization to fulfil their mandated responsibility of village development planning and management, and (c) to improve the village forest situation in order to sustainably manage and use the forests.

Important steps involved in strengthening village development planning are:

- Collection of basic socio-economic data;
- Identification and analysis of village level problems and opportunities;
- Preparation of the Village Development Plan including the Joint Village Agreement outlining the responsibilities for various activities like logging, forest rehabilitation, *etc.*;
- District level review and approval of the Village Development Plan;
- Implementation of the Village Development Plan including forestry development activities, for example, forest surveys, tree marking, logging, maintenance of logged-over areas, and forest rehabilitation which involves site preparation, nursery work, seedling preparation and transportation, planting and seedling maintenance.

Village development grants

Under the SUFORD (Sustainable Forest Management and Rural Development) Project, participating villages are provided with a Village Development Grant (VDG), of US $ 8,000 and 4,000 per village in the 1st and 2nd phase respectively. The objectives of the financial support are:

- To promote alternative means of income-generation resulting from village investments which could reduce the direct pressure on surrounding forests;

- To express goodwill or appreciation from the Government to those villagers for their participation and management of the production forests; and

- To give villages experience in administering revenues from forest management.

Provision of the Village Development Grant requires a number of activities including the formulation of the policies, specification of activities eligible for VDG support, type of activities prohibited, the process to be followed in decision making, process of selection of participant households, the procedure for implementation of various activities and so on.

Support to village finances management

Government also provides considerable support for fund mobilization including tracking the flow of funds from the central office to the villages, disbursement process, accessing funds and book keeping and auditing procedures. Specific attention is given to the following aspects.

- Fund management by villagers: village meeting for selection of Financial Management Unit of the Village Development Committee to be responsible for initial disbursements and procurement, contracting and payment to contractors and households and monitoring the payment of loans.

- Village Bank Accounting and Cash Box: village bank accounting opening and preparing for the use of the fund.

- Accounting and record keeping in relation to disbursement of VDG: the District staff may be helped and able to

  - Advise villagers on determining the most efficient and effective method for construction and procurement;

  - Assist villagers in managing construction and supervising works undertaken by a contractor; and

  - Report on project technical progress and problems, and certificate of completed works.

Provision of training and other technical assistance

The project staff also provide training and technical support to villagers on various issues such as:

- Suggesting or explaining to villagers that some of the VDG could be used to make loan to producer groups or individuals to establish small scale enterprises;

- Assisting (especially training and guiding) in conducting cost benefit analysis, activity models and gross margin budgets, business planning and accounting;

- Providing technical training and skills development and technical planning capacity development: conducting study tours to other similar ventures; and

- Developing linkage to markets, and/or developing an understanding of markets, market quality and design requirements.

Monitoring and evaluation

The monitoring and reporting of financial and physical progress of VDG activities will be conducted by the Village Development Committee (VDC). This includes the performance of the various activities and enterprises, including physical and financial progress and submission of reports to the District for each quarter.

Involvement of local stakeholders

A key factor in the success of the programme is the involvement of different stakeholders which include:

- District Governor's Office;
- Planning Office including the Statistics Office;
- Grassroots Construction and Rural Development Offices;
- Education and Health Offices;
- Lao Women Union;
- Lao National Front for Construction;
- Agriculture and Forestry Office.

Financial supports

The Government Forestry Development Projects focusing on forest conservation and rehabilitation for the three forest types in the whole economy are supported by the Forest Development Fund of the Lao PDR Government. Funds are also provided by the Lao – Swedish Upland Agriculture and Forestry Research Programme (LSUAFRP, Phase 1: 2004 – 2007, Phase 2: 2008 – 2011) for 4 provinces in northern Lao PDR.

The Sustainable Forest Management and Rural Development Project (SUFORD) supported and funded by the World Bank and the Government of Finland has implemented the project plan for the first phase from 2005 to 2008 and the Project has continued for the second phase during 2009 to 2011 in 5 provinces 16 districts 8 Production Forest Management Areas (PMA) in the economy. The main objectives of the SUFORD are to put priority natural production forests under sustainable forest management and to link such management with improvement of livelihood of villagers. At the village level, the SUFORD has two main components: ① production forest management, and ② village development. In the first phase —2005 to 2008 —the SUFORD has implemented the sustainable forest management plan in 4 provinces 18 districts involving 412 villages and 8 Production Forest Management Areas covering an area of 713,327 ha. For the second phase from 2009 to 2011, the SUFORD planned to continue and extend the project activities in the production forest management areas in 5 provinces 16 districts 714 villages participated, 8 Production Forest Management Areas (PMA).

Degraded Forest Rehabilitation Initiatives in Peninsular Malaysia 5

Jennifer AnakFrancis[1]

Abstract

Peninsular Malaysia is endowed with a large tract of tropical rainforest. This forest, rich in flora and fauna, is one of the most complex ecosystems on the earth. It is imperative that these invaluable forests are efficiently managed adopting the principles of sustainable forest management (SFM) that can enhance social, economic and environmental benefits. Forest rehabilitation, an important component of SFM, needs to be emphasized and effectively implemented to enhance productivity of the Permanent Reserved Forest (PRF). This paper highlighted the various initiatives on degraded forest rehabilitation by the Forestry Department of Peninsular Malaysia (FDPM), discussed the issues/challenges encountered and outlined the way forward to ensure success of various rehabilitation initiatives.

Introduction

Malaysia is a federation of 13 states and 3 federal territories with 11 of the states and the federal territories of Kuala Lumpur and Putrajaya located in Peninsular Malaysia, while the State of Sabah, State of Sarawak and Federal Territory of Labuan are located in the island of Borneo.

Located in the equatorial region, Malaysia is fortunate to be endowed with a large tract of

---

1. Assistant Director of Forestry Department, Peninsular Malaysia

rich tropical rainforests. These forests form an important life support system, contributing significantly to the livelihood of communities in terms of several forest products and environmental services.

Forests in Malaysia forms an integral component of one of the 17 "mega-biodiversity economies" of the world, consisting 15,000 species of flowering plants, 195 species of palms, 500 species of orchids, 1,159 species of ferns and fern allies, 400 species of fungus as well as 432 species of mosses. As for fauna, there are 286 species of mammals, 736 species of birds, 268 species of reptiles, 158 species of amphibians, 449 species of fresh water fishes and 150,000 species of invertebrates.

Conservation of such diversity is crucial for economic, social and environmental stability as each species, no matter how small, have important roles to play. In this context, Peninsular Malaysia has set aside sufficient forest areas and gazetted them as Permanent Reserved Forests (PRFs) to be managed in accordance with the principles of sustainable forest management (SFM) for the benefit of present and future generations.

The tropical rainforests in Peninsular Malaysia cover about 5.79 million hectares or 43.9 percent of the land area (Anon, 2006). These forests consist of unique and complex ecosystems which are home to the economy's rich flora and fauna. In Peninsular Malaysia, the major forest types consist of 4.4 million ha of dry inland forest which is the main forest cover, 0.24 million ha of peat swamp forest and 0.09 million ha of mangroves. Of the total forested area in Peninsular Malaysia, 4.89 million hectares are Permanent Reserved Forests (PRFs), 0.30 million hectares are State/Alienated Land Forests and 0.59 million hectares are National Parks/ Wildlife and Bird Sanctuaries. In the year of 2012, approximately 0.09 million ha of the PRFs are designated as production forests managed under sustainable forest management and the remaining are preserved as protection forests.

Besides its rich bio-diversity, these forests have contributed significantly to the socio-economic development of the economy. Further, these forests also play an important role in climate change adaptation and mitigation, safeguarding of water resources and prevention of natural disasters, such as floods and landslides. Hence, it is of paramount importance to set aside sufficient areas of PRF to be managed sustainably for the benefit of present and future generations.

An overview of degraded forest in peninsular malaysia

The International Tropical Timber Organization defines the forest degradation as the reduction of the capacity of a forest to produce goods and services. A degraded forest delivers a reduced supply of goods and services from a given site and maintains only limited biological diversity. It has lost the structure, function, species composition and/or productivity normally associated with the natural forest type expected at that area.

Degradation is an outcome of natural factors (forest fire, flood, landslide, *etc.*) or human interventions (encroachment, shifting cultivation, logging, *etc.*). Of all factors, the main cause of degradation in Peninsular Malaysia is forest fire. The peat swamp forests are particularly prone to fire which is exacerbated by climate change and at the same time contributes to climate change. In the first two months of 2014 itself, an area of approximately 0.50 million hectares of forested land have been affected by fire.

Apart from forest fire, shifting cultivation practiced by local communities for their livelihood is another factor contributing to forest degradation. In 2013, approximately 0.30 million ha of forested land in Peninsular Malaysia is identified as poor and degraded. According to the Asian Development Bank, about 0.05 million ha of forested land in Peninsular Malaysia has been degraded due to shifting cultivation practices in the year 1990. Figure 1 gives an indication of the distribution of degraded forests in Peninsular Malaysia.

However, proper planning and implementation of silvicultural treatments will enhance the potentials, especially through improving biodiversity, increasing the commercial value of forest products and improving soil fertility. Thus, the Forest Department of Peninsular Malaysia (FDPM) has produced its development plan for degraded forest in the Permanent Reserved Forests (PRFs). The objective of this plan is to restore and rehabilitate degraded area in PRF, to transform it into high-value forests increasing the production of wood and other environmental services. These forests also form an important habitat for biodiversity, food and medicinal plants and water catchment and contribute to climate stability. This development plan will cover up to 0.06 million ha of degraded forest in Peninsular Malaysia during 2011 and 2024.

FDPM perspective in rehabilitation of tropical forest towards sustainable forest management

According to the International Tropical Timber Organization (ITTO), sustainable forest management (SFM) has been defined by as "the process of managing forests to achieve one or more clearly specified objectives of management with regard to the production of continuous flow of desired forest products and services, without undue reduction of its inherent values and future productivity and without undue adverse impacts effects on the physical and social environment" (Anon., 1992). This is in line with the *National Forestry Policy 1978* (revised in 1992) which emphasizes that the Permanent Reserved Forests (PRFs) will be managed in accordance with the principles of SFM for maximization of the social, economic and environmental benefits of the nation. Relevant statements in the *National Forestry Policy 1978* (revised in 1992) with regard to the forest rehabilitation are:

- To manage the Permanent Forest Estate in order to maximize social, economic and environmental benefits for the nation and its people in accordance with the principles of sustainable management.

- To implement a planned programme of forest development through forest regeneration and rehabilitation operations in accordance with appropriate silvicultural practices, as well as the establishment of forest plantations of indigenous and exotic species to supplement timber supply from the natural forests.

Forest harvesting practices form one of the core areas that need to be addressed in achieving SFM. In this respect, FDPM has made systematic efforts directed towards research and development to formulate more environment-friendly harvesting technologies such as the use of reduced impact logging (RIL) so as to minimize the negative impact to the environment. In affirming the commitment to SFM, Malaysia has also developed a set of Malaysian Criteria, Indicators, Activities and Management (MC & I) in line with ITTO's Criteria & Indicators for monitoring and assessing SFM and also for the purposes of forest management certification to be undertaken at the forest management unit level. This is to ensure that there will be no poor and degraded forests in the area that is harvested.

One of the strategies is to utilize the Permanent Reserved Forest taking into account the inherent capability of the forests, giving due attention to forest regeneration and rehabilitation. FDPM has defined the Forest Regeneration and Rehabilitation as a coordinated programme of forest development through regeneration and rehabilitation operations based on appropriate silvicultural practices in order to achieve the maximum productivity from the PRF.

Degraded forest rehabilitation initiatives

Recognizing the importance of degraded forest rehabilitation in ensuring the sustainability of timber production and biodiversity conservation, FDPM has embarked on the following forest rehabilitation initiatives.

Selective Management System (SMS)

FDPM has been practising SMS in the dry inland forests, peat swamp forests and mangrove forests of PRF since 1978. It was implemented in Peninsular Malaysia due to the change in forest harvesting, mainly in the hill Dipterocarp forests, where conditions such as steep terrain and lower species richness do not favour a drastic opening of forest in one cut. The system focuses on a flexible timber harvesting regime with a cutting cycle, which varies depending on the forest types related to the differing pace of regeneration and growth. For example, the cutting cycle is 30 years for the dry inland forests, whereas for the mangrove forests it is 20 – 50 years.

Implementation of SMS involves three stages, namely inventory before logging (pre-felling), logging (felling) and inventory after logging (post-felling). Selective logging with prescribed cutting limit is regulated by the predetermined annual allowable cut which is revised every 5 years. In dry inland forests, under SMS, only 7 – 12 mature trees are felled in every hectare and 32 trees are left to form the next crop to be felled in the next rotation after 30 years. The system therefore guarantees economic viability and sustainability of log production for the next cutting cycle.

SMS also safeguards environmental quality and the maintenance of ecological balance. Harvesting of forests is also well coordinated and regulated to ensure compliance to environmental standards and full resource utilization. Obviously, the success of SMS will depend on the way that the forest harvesting practices are implemented.

Enrichment Planting Programme

The World Conservation Union defines the enrichment planting as the planting of desired tree species in modified natural forests or secondary forests or woodlands with the objective of creating a high forest dominated by desirable species (Anon, 2006). Enrichment planting enhances the productivity of an area by increasing the composition of high quality commercial timber species. In Peninsular Malaysia, enrichment planting is carried out in "poor forests" and "open areas". "Poor forests" are referred to as forests that have stocking of $153m^3$/ha while "open areas" are degraded forest areas or gaps created through activities such as shifting cultivation, forest encroachment and logging. Enrichment planting practices involve the planting of high-quality commercial timber species such as *Shorea leprosula* (Meranti tembaga), *Shorea parvifolia* (Meranti sarang punai), *Dryobalanops aromatica* (Kapur), *Hopea odarata* (Merawan siput jantan), *etc*. Two types of planting approaches are practiced, namely line planting and group planting. To date, a total 35,896 ha of PRFs in Peninsular Malaysia have been rehabilitated through enrichment planting (Table 1). This does not include another 101,069 ha of planted forests established in the PRF with exotic timber tree species (teak, pine, hevea, and acacia). Continued efforts are being made to identify "poor forests" and "open areas" within the PRFs for rehabilitation through enrichment planting.

Table 1 Enrichment planting sites planted with indigenous timber tree species in the permanent reserved forest (PRF) in Peninsular Malaysia during 1970 – 2012   ha

State	1970 – 2000	2001 – 2005	2006 – 2010	2011 – 2012	Total planted
Johor	1,804	648	44	168	2,664
Kedah	1,051	25	426	204	1,706
Kelantan	3,695	400	275	—	4,370
Melaka	399	—	76	—	475
N. Sembilan	958	210	199	66	1,433
Pahang	4,963	2,503	2,692	1,017	11,175
Perak	5,350	142	858	871	7,221
Perlis	150	—	10	100	260
Pulau Pinang	10	1	—	10	21
Selangor	3,661	336	280	203	4,480
Terengganu	1,779	60	—	252	2,091
Total	23,820	4,325	4,860	2,891	35,896

Retention of selected timber trees for fauna conservation

Tropical rainforests of Peninsular Malaysia are complex ecosystems with dynamic interdependency of flora and fauna forming the rich forest biodiversity. Realizing the important roles of fauna in maintaining the richness of forest biodiversity, FDPM has embarked on various in-situ and ex-situ flora and fauna conservation programmes with the cooperation of the Department of Wildlife and National Parks (DWNP). One of the outcomes of this programme is the ban on felling 32 timber species during logging operations in the PRFs as indicated in Table 2. These timber species produce fruits and seeds as food for many fauna such as primates, birds and squirrels.

Table 2 List of timber species retained for fauna conservation

Under-story species			
No	Scientific name	Local name	Consumption
1	*Aglaia* sp.	Bekak	Fruit (Primates & Birds)
2	*Archidendron bubalirum*	Kerdas	Fruit (Primates, Birds & Squirrels)
3	*Archidendron jiringa*	Jering	Fruit (Primates, Birds & Squirrels)
4	*Ardisia* sp.	Mata Pelanduk	Fruit (Primates, Birds)

(to be continued)

\multicolumn{4}{c}{Under-story species}
No	Scientific name	Local name	Consumption
5	*Artocarpus heterophyllus*	Nangka	Fruit (Primates, Birds & Squirrels)
6	*Artocarpus integer*	Cempedak	Fruit (Primates, Birds & Squirrels)
7	*Artocarpus rigidus*	Temponek	Fruit (Primates, Birds & Squirrels)
8	*Baccaurea maingayi*	Tampoi	Fruit (Primates & Birds)
9	*Baccaurea sumatrana*	Tampoi	Fruit (Primates & Birds)
10	*Barringtonia* sp.	Putat	Fruit (Birds)
11	*Boucea macrophyla*	Kundang Hutan	Fruit (Primates & Squirels)
12	*Durio* sp.	Durian	Fruit (Primates, Birds & Squirrels)
13	*Dysoxylum* sp.	Mersindok (Langsat Hutan)	Fruit (Primates & Birds)
14	*Eugenia* (*Syzygium*) sp.	Kelat Jambu Laut	Fruit (Primates & Birds)
15	*Garcinia artoviridis*	Asam Gelugor	Fruit (Primates & Birds)
16	*Mangifera indica*	Mangga	Fruit (Primates, Birds & Squirrels)
17	*Nephelium lappaceum*	Rambutan Hutan	Fruit (Primates & Birds)
18	*Sandoricum koetjape*	Sentul	Fruit (Primates & Birds)
\multicolumn{4}{c}{Over-story species}
No	Scientific name	Local name	Consumption
19	*Castanopsis* spp.	Berangan	Fruit (Primates & Squirrels)
20	*Dialium* sp.	Keranji	Fruit (Primates, Birds & Squirrels)
21	*Ficus* spp.	Ara	Fruit (Primates, Birds & Squirrels)
22	*Irvingia malayana*	Pauh	Fruit (Primates & Squirrels)
23	*Knema* sp.	Basong	Fruit (Primates & Birds)
24	*Koompasia excelsa*	Tualang	Depository of wild honey (Seed for Squirrels)
25	*Lithocarpus cyclophorus*	Mempening Gajah	Fruit (Primates)
26	*Mangifera longipetiolata*	Machang	Fruit (Primates & Birds)
27	*Myristica* sp.	Basong	Fruit (Primates, Birds & Squirrels)
28	*Parkia* sp.	Petai	Bean / Fruit (Primates, Birds & Squirrels)
29	*Podocarpus* sp.	Podo	Hill / Beach Conservation (Primates)
30	*Santiria laevigata*	Kedondong Gergaji Daun Licin	Fruit (Squirrels)
31	*Sterculia foetida*	Kelumpang Jari	Seeds (Birds & Squirrels)
32	*Sterculia parvifolia*	Kelumpang	Fruit (Squirrels)

Coastal rehabilitation and Conservation Programme

The tsunami tragedy in 2004 which involved 18 economies has highlighted the importance of mangrove forests in stabilizing the coastal environment. It also enhanced public awareness on conserving mangroves as part of damage mitigation. On 26 January 2005, the Honourable Prime Minister of Malaysia urged for increased efforts to conserve and protect the coastal ecosystems. This has enhanced the public support to the initiatives undertaken by FDPM in protecting the mangrove forests.

In 2005, FDPM had identified 8,416 ha of mangrove forest with different categories of risk, needing different silvilcultural treatments. Under the 9th Malaysia Plan, FDPM in collaboration with various government agencies and NGOs succeeded in planting and rehabilitating 2,088 ha of mangrove forest based on funding provided by the Federal Government. In the 10th Malaysia Plan, FDPM plans to rehabilitate 1,520 ha of mangrove forests which fall under the "high risk" category. This programme is intended to achieve the following objectives:

- To conserve natural coastline which serve as natural protection to minimize the damage caused by natural disaster and soil erosion;
- To create buffer zones to withstand the high waves and strong wind as well as prevent environmental pollution;
- To restore coastal habitat that serve as corridors and enrich the biodiversity; and
- To improve environmental quality and aesthetic value as a tourist attraction.

During 2005 and 2012, an area of 2384 ha of land have been successfully planted with about 6.3 million trees. This includes over 5.9 million mangrove tree species (*Rhizosphere apiculate, Rhizosphere mucronata, Avicennia alba, etc.*), 194 thousand Rhu trees (*Casuarina equisetifolia*) and 144 thousand trees from other species (*Calophyllum inophyllum, Xylocarpus moluccensis, Nypa fruticans, etc.*).

Tree planting through public awareness programmes

Under the various public awareness programmes in Malaysia, there are a number of tree-planting campaigns/activities organized by FDPM with the participation of various government agencies, NGOs, private companies, school children and general public. Some of the important public awareness programmes involving tree-planting activities are as follows.

26 Million Trees -planting Campaign

Tree planting is regarded as one of the most effective initiative to preserve, conserve and protect the Mother Nature and ultimately the Earth. In conjunction with the World Earth Day on 22 April 2010, the Ministry of Natural Resources and Environment, Malaysia launched the 26 Million Trees-planting Campaign in Putrajaya carrying the theme "Green the Earth: One Citizen, One Tree". This campaign is in line with Malaysia's commitment during the Rio Summit 1992 of greening 50 percent of the economy's land area. The campaign was also one of the strategies to meet Malaysia's commitment to voluntarily cut 40 percent of the economy's carbon emission by 2020. A total of 26 million trees was targeted to be planted during the period 2010 – 2014 over approximately 13,066 ha throughout the economy. As of May 2014, a total of 20.9 million trees have been planted over an area of about 22,241 ha. Various agencies are involved in this tree-planting campaign and FDPM is responsible for monitoring the progress of planting under this campaign.

The International Day of Forest

The International Day of Forest is celebrated around the world on 21 March to commemorate the contribution and value of forests and forestry to the community. Besides, it is also to raise people's awareness about the importance of all types of forests. FDPM usually include tree planting as one of the activities in this programme. The theme for the International Day of Forest 2014 is "Forests for Community Livelihood".

International Year of Forests 2011 (IYF 2011)

IYF 2011 was launched on the 9th United Nations Forum on Forests on 2 Feb 2011 in New York with the theme "Forests for People". In Malaysia, the IYF 2011 was jointly launched by the Chief Minister of Johor State and the Minister of Natural Resources and Environment Malaysia in conjunction with the national level World Forestry Day celebration held in Nusajaya, Johor on 21 March 2011. A total of 1,000 trees were planted.

The Malaysia Book of Records—Largest Paya Bakau Tree-planting Event 2014

In conjunction with the Ship for South-East Asian Youth Programme (SSEAYP) annual assembly, the Institute of Foresters Malaysia (IRIM) in collaboration with FDPM and other agencies including NGOs organized the Largest Paya Bakau Tree-planting Event in May 2014 in Langkawi Island. Officiated by the former Prime Minister of Malaysia, Tun Abdullah Ahmad Badawi and his wife Tun Jeanne Abdullah, 20,200 mangrove trees were planted successfully. The objective of this programme was to help create a natural barrier and be a natural nursery for small fish and other aquatic life. This programme broke the previous record of planting 11,111 mangrove seedlings in 2011.

Issues / challenges and the way forward

Forest rehabilitation involves manpower and regular funding. As years go by, the cost of forest rehabilitation escalates and every effort is being made by FDPM to cover a larger area to fulfill Malaysia's commitment of greening 50 percent of the economy's land area and adhering with the SFM concept. Trees planted needs to be systematically maintained to ensure good survival and production of quality timber. Collaboration with various Government and its related agencies as well as NGOs are required to overcome funding issues. Under the Federal Constitution, forestry and land are the State matters. Therefore, full commitment and strong financial support from the various the State Governments are essential for the successful implementation of SFM and in particular the implementation of forest rehabilitation programmes by FDPM.

The regeneration of forest depends on the climatic, edaphic and biotic factors of an area (Symington, 2004). Each forest is unique and requires different silvilcultural treatment to achieve the maximum productivity. Failure to understand the dynamics of forest stand structure, in particular the effect of soil and climatic conditions on natural regeneration of forest, may lead to inappropriate silvilcultural prescriptions. More research on the dynamics of tropical rainforests needs to be carried out to ensure the successful implementation of forest rehabilitation programmes.

Landscape level information on the species composition and stocking of flora and fauna in the various forest types are essential for the rehabilitation of tropical rainforests. Currently such information is limited and insufficient. Therefore, further research needs to be carried out to develop practical and cost-effective techniques especially for forest inventory. Besides, to ensure biodiversity conservation under sustainable forest management, long-term monitoring and research and development (R & D) programmes are required to test the validity of the concepts and approaches to forest rehabilitation. However, this requires long-term financial, institutional, logistic and intellectual commitments. This also requires building networks and strengthening cooperation between research institutions and universities at the national and international levels.

Adverse human activities in the forest, which include poaching/hunting and illegal logging, are threats to conservation of biodiversity. Reduction in soil organic matter commonly follows conversion of natural forest to other forms of land use including plantations. For example, there has been rapid expansion of oil palm plantation in Malaysia, and most of these involved the clearance of rain forests and destruction of peat lands.

There is no doubt that the awareness among local community towards the importance of forests and biodiversity has increased. However, there are still individuals who are not yet to fully understand the importance of this matter. Therefore, community awareness programmes are essential to educate communities, loggers and forestry department staff on the beneficial aspects of biodiversity conservation and on the need to adopt SFM.

Conclusions

Forest rehabilitation is essential for the enhancement of timber productivity, sustainable supply of timber, conservation of biodiversity and stability of environment. It is an important component in the Sustainable Forest Management practices implemented by FDPM. Recognizing this, FDPM has taken up a number of initiatives to implement various forest rehabilitation programmes. To enhance the successful implementation of these programmes, R & D needs to be carried out by forest research institutions and universities particularly on the dynamics of tropical rainforests at the landscape level. However, forest rehabilitation and R & D require manpower and secure funding. As forestry and land are the State matters, it is imperative that the state governments are committed to the implementation of SFM and provide sufficient funding for the various SFM activities, in particular forest rehabilitation programmes. Although the current level of forested land is expected to decline by 2020, the total forest areas under the PRFs in Malaysia at the end of 2020 is expected to increase with forest rehabilitation practices. This will contribute to meeting the four global objectives on forests of the *Non-legally Binding Instrument on All Types of Forests*, especially Global Objective 1, among others, in reversing the loss of forest cover worldwide through sustainable forest management.

Acknowledgements

The author would like to thank the Director General of Forestry, Peninsular Malaysia for giving the opportunity to present this paper.

References

Anon. 1992. ITTO Guidelines for the Sustainable Management of Natural Tropical Forests. (Series Number 1). Yokohama, Japan; International Timber Trade Organization.

Anon. 2006. Guidelines for the Conservation and Sustainable Use of Biodiversity in Tropical Timber Production Forests. The World Conservation Union.

Anon. 2011.Selected Indicators for Agriculture, Crops and Livestock 2006 – 2010. Department of Statistics Malaysia.

Anon. 2012. Forestry Statistics Peninsular Malaysia. Forestry Department Peninsular Malaysia.

Anon. 2013. Development Plan for Degraded Forest in the Permanent Reserved Forest (PFRs). Forestry Department, Peninsular Malaysia.

Symington CF. 2004. Ecological distribution of the Dipterocarps in the Malay Peninsula: Foresters' Manual of Dipterocarpaceae. In: Ashton P.S. & Appanah (rev.). S. Malayan Forest Records No. 16. Kuala Lumpur: Forest Research Institute Malaysia & Malaysian Nature Society, 14–32.

Degraded Forest Rehabilitation and Sustainable Forest Management in Myanmar

Chaw Chaw Sein[1]

Abstract

Myanmar is a forest rich economy with a long history of scientific management. The paper outlined the current state of resources, the evolution of forest management and the changes in policies and legislation which enabled the forest sector to respond to the changing needs of society. Policies and legislation had attempted to give increasing thrust on environmental protection and meeting the needs of the local communities. Considering the vastness of the economy and the diversity of conditions, a wide array of approaches have been developed to cater to the differing needs and conditions.

Introduction

Myanmar is a continental Southeast Asian economy with a geographical area of 676,577 km² sharing its boundary with China in the north, Lao PDR and Thailand in the east, and Bangladesh and India in the west. The Andaman Sea and the Bay of Bengal edge the Myanmar coast in the south and the west. Topographically, Myanmar can be divided into three regions: the western hills, the central valley, and the eastern hills. The general profile of the economy rises from the sea level along the southern coasts to the snow-capped mountains reaching the highest elevation of around 6,000 m in the northern tip of the economy. Myanmar has a typical tropical monsoon climate and the economy can generally be described

---

1. Staff Officer of Forest Research Institute, Yezin

as hilly and mountainous. It is drained by many river systems most of which flow from the north to the south. As a result of the great variations in rainfall, temperature and topography, there are many forest types in Myanmar. Tropical evergreen forests occur in the high rainfall zone especially in the southern part of the economy. Hill and moist forests are found in the eastern, northern and western zones where the elevation exceeds 900 m. The vegetation changes into deciduous, then into dry forests towards the middle of the economy as a result of low rainfall. Mangrove forests are the characteristic vegetation in the coastal areas.

Forest resources and forest management in myanmar

Forest resources

Myanmar has a forest area of 35.38 million ha or about 52 percent of the land area. The forests are highly variable from scrubby and thorny growth in central Myanmar to the evergreen dipterocarps forests (Table 1). Mixed deciduous forests accounting for about 37 percent of the forests are the most prominent and probably the most important in economic terms.

Table 1 Forest types in Myanmar

Type of forest	Area (1000 ha)	Percent of total forest area (%)
Tidal Forest, Beach and Dune Forest and Swamp Forest	1,375	4
Tropical Evergreen Forest	5,500	16
Mixed Deciduous Forest	13,407	37
Dry Forest	3,483	10
Deciduous Dipterocarp Forest	1,719	5
Hill and Temperate Evergreen Forest	8,939	25
Scrub Land	998	3
Total Forest Area	35,421	100

Of the known 7,000 plant species in the Myanmar forests, 1,071 are endemic. Out of the 2,088 tree species, 85 species have been recognized and accepted as producing multiple-use timber of premium quality. Studies on the properties and utilization of the lesser-used timber species (LUS) are being carried out, and their use is extensively promoted. The objective is to increase their commercial use is and thus to reduce the pressure on the premium-quality timber species.

Permanent forest estate in Myanmar

Forest land is owned by the State and is legally classified as reserved forests, protected areas and public protected or unclassified forests. So far more about 20 million ha of the forest area (approximately 30 percent of the total economy's land area) have been classified as permanent forest estate (PFE), which comprises of conservation reserves (i.e., protected areas), production forests (i.e., forest reserves), and protected public forests (Table 2). The number and extent of protected areas have been steadily increasing since the early 1920s (Figure 1).

Table 2 Permanent forest estate in 2010

Legal classification	Area (km^2)	Percent of total land area (%)
Permanent Forest Estate	197,899	30.7
Forest Reserves	121,843	18.0
Protected Public Forest	40,950	6.0
Protected Areas System	35,107	6.7

Source: Planning and Statistic Division, Forest Department, 2011.

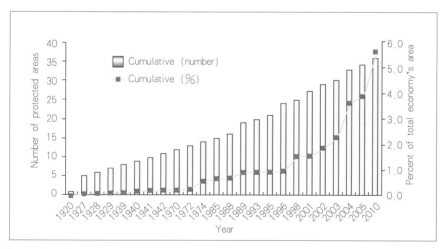

Figure 1 Cumulative number of protected areas between 1920 and 2010

Forest management

Myanmar has a long history of scientific forest management dating back to 1856 when the system of management based on working plans was introduced to manage the teak forests. Forest management in the economy was being guided by 33 plans; however there was some discontinuity on account of the delay in the timely revision of these working plans. In 1996, the Forest Department launched a special operation to update and reformulate the working plans by incorporating the concepts of sustainable forest management (SFM). Working plans have been renamed as forest management plans, and these give emphasis not only on timber production but also on the production of non-wood forest products, biodiversity conservation, well-being of local communities, in particular forest-dependent people, all of which are integral to the practice of sustainable forest management. By the end of 1998, forest management plans have been prepared for 62 districts covering the entire forests in the economy.

Myanmar Selection System

Scientific management of Myanmar forests started in 1856 by Dr. Brandis. He introduced a system of selective logging of teak trees above a certain girth limit based on observations about the rate of growth. This Exploitation-cum-cultural system, known as the Myanmar Selection System (MSS) has been the principal silvicultural system followed in the management of Myanmar's natural forests, including for species other than teak. The MSS involves the determination of the annual allowable cut (AAC) for teak and other hardwood species based on a felling cycle of 30 years and a specified exploitable size, selective marking and felling of trees (with girdling of teak trees prior to felling), felling of less-valuable trees interfering with the growth of teak and the thinning of congested teak stands. The strict adoption of MSS enables long-term sustainable wood production, provided the exploitation is limited to the AAC. However illegal logging tends to undermine sustainability.

Myanmar has been undertaking national forest inventory yearly with the aid of remote sensing and geographic information system taking advantage of landsat imageries in each and every state and division alternatively. This has helped monitor the state of forests and adjust the yield.

Policy, legislation and institutional arrangements

Forest policy

In view of the importance of the forestry sector to national socio-economic development, and the ensurance of ecological balance and environmental stability, the Myanmar forest policy has been formulated in a holistic and balanced manner within the overall context of the environment and sustainable development policies and strategies as also taking full cognizance of the forestry principles adopted by the United Nations Conference on Environment and Development, 1992.The policy formulated in 1995 has identified 6 imperatives which the Government must give the highest priority in order to achieve broader national development goals and objectives. They are:

a. Protection of soil, water, wildlife, biodiversity and environment;

b. Sustainability of forest resources to ensure perpetual supply of both tangible and intangible benefits from the forests for the present and future generations;

c. Basic needs of the people for fuel, shelter, food and recreation;

d. Efficiency in harnessing, in a socio-environmentally friendly manner, the full economic potential of the forest resources;

e. Participation of the people in the conservation and utilization of the forests; and

f. Public awareness about the vital role of the forests in the well-being and socio-economic development of the nation.

National Environmental Policy

The National Commission for Environmental Affairs (NCEA) was established in 1990 to coordinate environmental matters across the different ministries, to develop the National Environmental Policy and to liaise with foreign economies and non-government organizations regarding environmental matters. Myanmar's National Environmental Policy was formulated in 1994 integrating environmental considerations in social and economic development. The primary responsibility for implementing the national policy on nature and biodiversity conservation in Myanmar is vested with the Ministry of Environmental Conservation and Forestry (MOECAF) but other ministries such as the Ministy of Agriculture and Irrigation (MOAI), Ministry of Livestock and Fisheries (MOLF), *etc.*, share the common responsibility and accountability for biodiversity conservation.

Legal framework

For providing the legal backing to implement the 1994 National Environmental Policy, Myanmar has formulated various laws, rules and regulations. The *Protection of Wildlife, Wild Plants and Conservation of Natural Area Law (1994)* mandates protection of wild flora and fauna and their habitats and representative ecosystems. The *List of Protected Species (1996)* makes provision for various levels of protection for wild plants and animals as indicated below.

a. Completely protected species may not be hunted except for scientific purposes under special license.

b. Protected species may be hunted but only with special permission.

c. Seasonally protected species are subjected to traditional subsistence hunting by rural communities only during the open (i.e., non-breeding) season (Annex III).

In addition, the *Forest Law (1992)* protects forest resources and gives priority to link forest management to social and environmental considerations. Further, in 2012, Myanmar has enacted the *Environmental Law* to implement the National Environment Policy to ensure that development activities have minimum negative environmental impacts.

Myanmar Agenda 21

In compliance with Myanmar's National Environmental Policy, *Myanmar Agenda 21* was developed in 1997 and it was a collaborative effort involving various government agencies in order to strive for the sustainable development of the economy. The formulation of *Myanmar Agenda 21* is an important step in the process of achieving sustainable development and is intended to serve as a framework for integrating environmental considerations in the national development plans as well as the sectoral and regional development programs. This document is the expression of the political commitment of the Government to sustainable development. *Myanmar Agenda 21* underpins the importance of nature and biodiversity conservation, particularly giving attention to protected area planning and management and biodiversity conservation.

30-year National Forest Master Plan (NFMP)

Forest resources have played an important role not only in the socio-economic development but also in ecosystem stability and biodiversity conservation in Myanmar. Taking this into account, MOECAF has formulated a National Forest Master Plan (NFMP) with a time horizon of 30 years from 2001 – 2002 to 2030 – 2031. The NFMP provides a comprehensive framework focusing on a number of strategic areas such as (a) management of natural resources, (b) establishment of forest plantations, (c) establishment of community forests, (d) growing trees in homestead and non-forested areas, and (e) promotion of wood-based industry value-added forest products. It is envisaged that by 2030 – 2031 the extent of PFEs would be increased to 40 percent of the economy's land area.

Dry Zone Greening Action Plan

With the aim to conserve and promote the rehabilitation of dry land ecosystems of the central Dry Zone, a comprehensive 30-year Dry Zone Greening Action Plan, divided into 5-year medium-term plans, has been formulated. The Greening Action Plan involves the establishment of forest plantations in degraded areas using suitable native and exotic tree species. In addition, the degraded and remnant natural forests are improved through natural regeneration and enrichment planting.

National biosafety framework

From May 2004 to November 2006, a project of "Development of National Biosafety Framework Project, Myanmar" was implemented with technical and financial assistance from the United Nations Environment Program (UNEP) and the Global Environment Facility (GEF). The objective of the Project was to support biotechnology development while guarding the national biodiversity in a sustainable way as well as ensuring human health. At the completion of the Project, a *Draft on National Biosafety Framework* and a *Draft Law on Biosafety* had been prepared and these were under consideration by higher authorities.

National Biodiversity Strategy and Action Plan

In cooperation with UNEP/GEF, the Forest Department has formulated the National Biodiversity Strategy and Action Plan (NBSAP). The NBSAP, adopted by the Government in May 2012, aims to provide a strategic planning framework for the effective and efficient conservation and management of biodiversity. The specific objectives of NBSAP are (a) to set the priority for conservation investment in biodiversity management, and (b) to develop a range of options for addressing the issue of biodiversity conservation. The development of NBSAP will facilitate the sustainable use of biological resources and help comply with Myanmar's obligations to the *Convention on Biological Diversity*. The NBSAP is expected to provide a comprehensive framework for planning biodiversity conservation, management and utilization in a sustainable manner as well as to protect Myanmar's rich biodiversity.

National Environmental Conservation Committee

In November 2004, Myanmar established the National Environmental Conservation Committee (NECC) with the objective of promoting environmental conservation and sustainable development of the economy and thus to give environmental matters a priority. NECC aims to consolidate the environmental conservation activities at the local and national levels. It is chaired by the Union Minister of the MOECAF. In April 2011, NECC was reformed by including 21 members from 19 ministries. Sub-committees have also been formed for each eco-region. Important functions of the NECC are:

- To address the environmental problems induced by unsustainable land use;
- To address the environmental problems in rivers and wetland areas;
- To implement environmental conservation activities in industrial zones and civil areas;
- To develop policies, principles, rules and regulations for environmental matters; and
- To strengthen environmental awareness.

NECC has been monitoring the implementation of environment conservation activities by its members in the economy and is giving necessary advice to achieve the objectives.

Institutional arrangements

The Ministry of Environmental Conservation and Forestry is the primary institution responsible for forestry and environment and there are 6 departments under the Ministry assigned with different responsibilities as indicated in the organigram below (Figure 2).

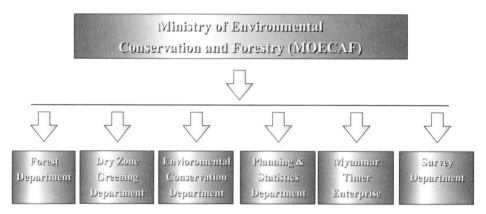

Figure 2 Structure of the Ministry of Enviromental Conservation and Forestry

Major drivers of forest degradation

As in other economies, Myanmar is also facing problems of deforestation and forest degradation, the major drivers of which are overexploitation, illegal logging, shifting cultivation, expansion of agricultural land, fuelwood collection, and urbanization.

Rehabilitation and reforestation activities

The rehabilitation and reforestation program in Myanmar commenced in the early 1960s and large-scale plantation forestry in the 1980s. The Forest Department mainly establishes four types of plantations: commercial plantations, industrial plantations, watershed plantations for rehabilitation of degraded watershed areas and local supply plantations to enhance wood fuel supply to local communities.

Combating desertification in the Central Dry Zone

The Dry Zone Greening Department (DZGD) was formed in 1997 with the special task of preventing desertification in the Dry Zone of the Central Myanmar. In pursuance of this objective, DZGD is focusing on (a) the establishment of forest plantations for local supply and greening purpose, (b) protection and conservation of the remaining natural forests, (c) wood-fuel substitution, and (d) development of water resources. A number of international organizations and companies are also contributing to the environmental restoration in the dry zone through afforestation efforts, including establishment of forest plantations.

Greening activities of the Bago Yoma Range

In the Bago Yoma Range, teak (*Tectona grandis* Linn. f) forests dominate and deforestation and degradation take place due to its easier accessibility compared with other parts of the economy. The Forestry Department (FD) has therefore been implementing the Bago Yoma Greening Project covering 6 districts with a total area of about 50,700 km^2 with the objective of restoring and rehabilitating the degraded forests. The major activities under the Project are conservation and protection of natural forests, enrichment planting, natural regeneration and establishment of plantations. Other restoration components include the establishment of community forests, provision of extension services, supply of water, establishments of seed production area and implementation of research activities.

Conservation of coastal and mangrove forests

Due to extensive use of the mangroves for production of charcoal and timber as well as encroachment for cultivation, the extent of mangroves has declined from 386,000 ha in the 1990s to about half by 2002. Most of the mangrove forests suffer from various levels of degradation, rendering the rehabilitation of these forests a crucial issue. As part of its regular activity, the FD is implementing mangrove rehabilitation involving improvement felling to encourage regeneration, plantation establishment in depleted areas and abandoned paddy fields, and community forestry. In cooperation with international organizations, conservation and rehabilitation projects have been also carried out.

Forestry development in border areas

Border areas of Myanmar are mostly rugged and mountainous, particularly in the northwest, north and the east. The Development for the Progress of Border Areas and National Races Department (DPBANRD) in collaboration with other related governmental and non-governmental bodies has been undertaking social, economic and environmental development tasks.

Establishing a protected area system and biodiversity conservation

Myanmar has been a member economy of the International Center for Integrated Mountain Development (ICIMOD) and a signatory of the *Convention on Biodiversity Conservation* (CBD) since the early 1990s. Up to now, 32 wildlife sanctuaries and 2 national parks, constituting about 7.3 percent (49,456.46 km^2) have been formed and managed. The FD envisages the expansion of the protected areas so that they will eventually account for 10 percent of the land area.

Adoption of criteria and indicators (C & I) for SFM

Identification of Myanmar's C & I for SFM at both national and forest management unit (FMU) levels was completed in 1999, and is based on International Tropical Timber Organization (ITTO)'s C & I of 1998. It contains 7 criteria each for both national and FMU levels. There are 78 indicators and 257 required activities at the national level, and 73 indicators and 217 activities at the FMU level, together with standards of performance for each activity. The FD has been testing the adequacy and application of Myanmar's C & I at FMU level for further improvement. The MOECAF and the Forest Resources and Environment Development Association (FREDA), a forestry-related non-governmental organizations (NGOs) in Myanmar, took this initiative with the financial and technical support from Japan Overseas Forestry Consultants Association (JOFCA).

Model demonstration forests

The FD has established two model forests in the Bago Yoma region. The Japan International Forestry Promotion and Cooperation Center (JIFPRO), the Japan Overseas Forestry Consultants Association (JOFCA) and some of the NGOs from Japan have been cooperating with the FD in developing these model forests to demonstrate forest certification.

Timber certification

The Timber Certification Committee (TCC) was formed in August 1998 by the MOECAF, and since then, it has been establishing links with other timber certification bodies on a bilateral basis. Myanmar TCC has links with the National Timber Certification Council (NTCC) of Malaysia and the Eco-labeling Institute of Indonesia (LEI). Myanmar's TCC is now spearheading the development of a timber certification scheme appropriate to Myanmar's forest management system. Myanmar's C & I is the basis for developing a timber certification checklist at the FMU level. Nowadays, MOECAF with the assistance from European Union has been processing an agreement for certifying the legal timber.

Forest-harvesting code

MOECAF has also developed the *National Code of Forest Harvesting Practices in Myanmar* in 1999 with financial and technical assistance from the Food and Agriculture Organization of the United Nations (FAO). A number of training courses have been provided to the staff of the Extraction Department of MTE to promote the implementation of improved logging practices.

Nation-wide tree-planting campaign

Realizing the direct and indirect benefits of trees in mitigating climate change, the Forest Department has launched a nation-wide tree-planting programme since 1997 – 1998 with the objective of raising public awareness and greening the non-forest area to improve the provision of environmental services. Some of these services have direct positive impacts of fulfilling basic needs of local people.

Management of watershed areas

Restoration of watershed areas especially in the catchment of important dams started early. Excessive removal of vegetative cover, practice of slash and burn system including the shortening of fallow period and overgrazing are the major causes of watershed degradation. The FD has established a number of watershed protection plantations. In addition, the United Nations Development Programme (UNDP)/FAO supported the Pilot Watershed Management Project for the Kinda Dam, Inlay Lake (Shan State) and Phu-gyi Dam Watershed has helped strengthen the FD initiatives. Since the completion of the UNDP/FAO project, the FD is sustaining the various project activities.

Stabilizing shifting cultivation

It is evident that shifting cultivation is one of the major causes of forest depletion and degradation. To address this, a national level multi-sectoral program of highlands reclamation has been developed. The program aims to support traditional land use systems, customary rights and cultural values. In cooperation with other sectors, the FD has been implementing activities such as (a) community forestry based on agroforestry systems, (b) provision of improved technologies, complementing traditional forest-related local knowledge, (c) recruiting shifting cultivators into routine forestry operations such as plantation establishment, (d) enhancing income-generating opportunities, and (e) implementation of awareness-raising campaigns and extension services.

Controlling illegal logging

Illegal logging today is a widespread problem and is threatening forest management in most developing economies. The following measures are being undertaken in Myanmar to reduce illegal logging and to eliminate it completely:

- Strict enforcement of the existing forest law, rules and regulations;
- Setting up checkpoints along the main shipping routes across the economy;
- Inspection of logging operations to ensure that they are carried out in accordance with the procedures and prescribed rules and regulations;
- Implementation of an incentive scheme for the staff and those who are actively engaged in protecting illegal logging;
- Forming a partnership with the institutions concerned and local communities to prevent

and detect illegal logging; and

- Cooperation and coordination with the neighboring economies in combating illegal logging along the borders.

Community forestry

The FD has issued the *Community Forestry Instructions* (CFI) in 1995 so as to develop community participation in forest management in Myanmar. CFI allows local communities to be involved in developing management plans, protection, conservation and restoration of forests, and encourages the utilization of forest products and Non-Timber Forest Products (NTFPs) particularly in the vicinities of their settlements. The CFI allows for a 30-year land use, ownership rights and disposal of products from community forest under the guidance of the FD.

Education, training and research

The University of Forestry located in Yezin is the only institution in the economy giving education relating to forestry and environmental sciences. In addition, there are 3 centers providing training to staff of the forestry sector as well as private individuals and communities. Another three training centers under the MTE are conducting courses on timber extraction and downstream processing. Site-level restoration research has focused on assessment of stand characteristics and existing natural regeneration, and enrichment planting. Research programs and activities of the Forest Research Institute are merely formulated in line with forest policy at the national level in addition to regional and global concerns. To adopt more appropriate training and research program, it is necessary to improve human resource and capacity management.

Forestry extension

In order to promote forestry and environmental service, the Extension Division was formed under the FD with the objectives of (a) publishing and disseminating forestry-related articles in government newspaper, (b) publishing the Myanmar Forestry Journal, (c) broadcasting videos, and films relating to conservation of natural resources and environment on television, (d) organizing exhibitions and other events at national events, and (e) enhancing public awareness and participation.

7. Sustainable Forest Rehabilitation and Management for the Conservation of Trans-Boundary Ecological Security in Montane Mainland Southeast Asia: Pilot Demonstration Project in Myanmar

Chaw Chaw Sein[1]

Abstract

Myanmar is participating in the APFNet's multi-economy project on the Sustainable Forest Rehabilitation and Management for the Conservation of Trans-boundary Ecological Security in Montane Mainland Southeast Asia and the paper described the activities being undertaken in the demonstration area in Myanmar. The area is characterized by diverse land uses including shifting cultivation, home gardens and natural forests. The paper described the implementation of different participatory rehabilitation approaches appropriate to the diverse land uses.

Introduction

The target village for implementing participatory land use plan is located near the Nyuak-Htauk reserved forest and it includes residential areas, private farm lands, cultivation-shifting lands in reserved forests and un-classed forests, spiritual forests, spring water forests and forests where villagers collect fuelwood for domestic consumption. Different forest rehabilitation strategies have been designed for each land use taking into account the current state of land use determined on the basis of a baseline assessment. During this reporting period, sustainable forest rehabilitation plan for home gardens, cultivation-shifting lands, spiritual forests and spring water resources were formulated adopting a participatory process

---

1. Staff Officer of Forest Research Institute, Yezin, Myanmar

involving local communities, government agencies and scientists. The following sections elaborate the management plan for forest rehabilitation.

Forest rehabilitation in community-based forest management

Forest rehabilitation plan was prepared based on a review of literature on best practices, field survey, local perceptions and concerns, involvement of local officials and local forest department staff.

Establishment of community forests in cultivation-shifting areas

The pilot demonstration site is located in the reserved forest where local communities have been practising shifting cultivation since 1988. Of the total area of 24.3 ha, 8.41 ha is affected by shifting cultivation while the remaining 15.89 ha consists of degraded secondary forests with teak (*Tectona grandis*) and some hardwoods species such as thit-ya (*Shorea obtusa*) and in-gyin (*Shorea siamensis*). Due to increase in population and industrialization of agriculture, shifting cultivation—taung-ya—is likely to change into sedentary agriculture. The most common crops planted in the taung-ya include maize, groundnut, sesame and paddy (Figure 1). Maize has become a popular commercial crop for the local community on account of marketability and the ease of cultivation in comparison with other crops. The industrialization of agriculture with intensive use of inputs has increased productivity and farmers' income; but it also has some negative impacts, especially by way of reducing agro-ecosystem diversity and increasing soil erosion. Community forestry became the thrust area for developing land use models, especially for tackling the twin problem of degradation and livelihood insecurity.

Figure 1   Crops planted in the taung-ya

Management plan for community forestry with agroforestry

Community forest user group was organized with 9 households who are doing Taung-ya in the 8.41 ha of cultivated land. According to the state of vegetation and present land use, two management strategies were designed under community forestry, i.e. agroforestry, and conservation of existing natural forests. Each of the households would manage both agroforestry as also natural forest plots based on a management plan prepared through a collaborative process involving user group members, authorities, elder people and scientists.

The preparation of management plan involved the following steps.

- Gathering community information and identifying village needs through village meeting and informal interview by scientists;
- Presenting the concept of community forestry and agroforestry in the village meeting;
- Sharing economic outcomes of agroforestry using examples from other user groups;
- Facilitating local communities to develop preferable agroforestry design including listing the preferred species to be planted;
- Getting concurrence and acceptance of the agroforestry design in the meeting;
- Discussion about conservation of existing natural forests;
- Building consensus among user members for the conservation of natural forests with technical support from scientists;
- Site visit to Wundwin Township community forest to study the present rehabilitation practices of agroforestry;
- Establishment of nursery and preparation of planting materials;
- Demonstration of tree management practices to user group members such as planting trees and improvement felling by local forest department (FD) staff and scientists;
- Identification of tree vegetation in community forest areas and home gardens to measure the success of forest rehabilitation; and
- Raising awareness of user group members in sustainable forest rehabilitation. Training was also provided in related areas like compost making to local communities with the aim of raising awareness about soil management practices.

The layout design for agroforestry varies among members based on their species preferences, availability of land, traditional knowledge and scientific knowledge on agroforestry practices.

For intercropping, 12 ft.[①] ×12 ft. wide inter-row spacing is used for households who have above 10 acres[②] of cultivation-shifting area as shown in Figure 2. In the case of user group members who have limited cultivation-shifting area, 12 ft. × 24 ft. spacing or 36 ft. × 36 ft. spacing (Figure 3) is to be followed.

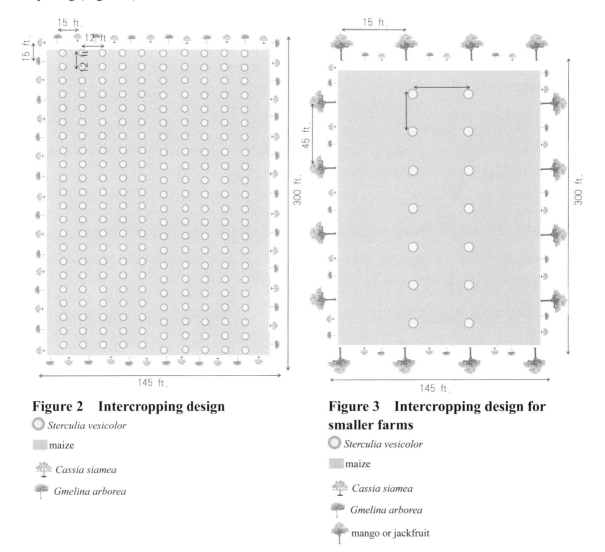

Figure 2   Intercropping design
○ *Sterculia vesicolor*
▨ maize
※ *Cassia siamea*
♠ *Gmelina arborea*

Figure 3   Intercropping design for smaller farms
○ *Sterculia vesicolor*
▨ maize
※ *Cassia siamea*
♠ *Gmelina arborea*
♣ mango or jackfruit

During last year's planting season, 6 members have planted about 2550 seedlings of Shaw Phyu (*Sterculia versicolor*) in their individual farm land at a spacing of 12 ft. × 12 ft. as shown in Figures 4 and 5. Local people prefer to plant shaw-phyu trees on account of their fast growth and the higher economic returns within a short period. Other trees proposed to be planted in the following years are mango (*Mangifera indica*), jackfruit (improved varieties

---

① 1 feet (ft.) = 0.3048 m. The same below.
② 1 acre = 4046.86 m². The same below.

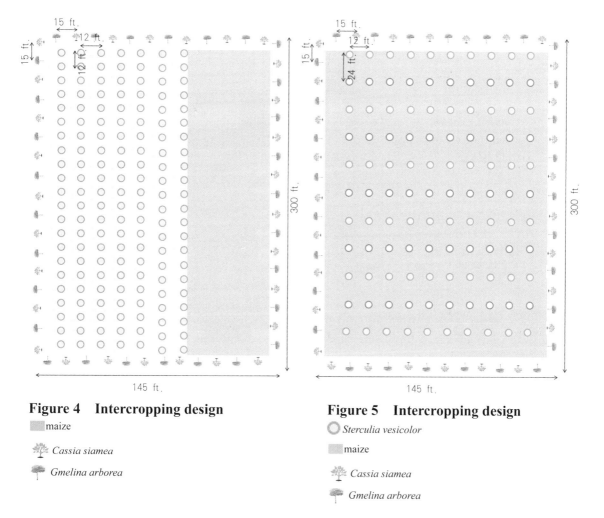

Figure 4 Intercropping design
maize
Cassia siamea
Gmelina arborea

Figure 5 Intercropping design
Sterculia vesicolor
maize
Cassia siamea
Gmelina arborea

which fruit in three years), lemon and danyin. These fruit trees are planted to provide subsistence needs as well as to generate additional income from the farm land. Most of these fruit yielding trees help improve the soil conditions.

The user group members usually grow maize, groundnut or rice in the interspaces. Even though the shifting cultivation fallow system has largely disappeared, some farmers still practise the traditional knowledge-based rotational cropping system to reduce soil degradation. For the first year, they plant groundnut or rice (mostly groundnut) and maize is then cultivated for three consecutive years. After that, they plant again groundnut. Figure 4 shows that local communities plant groundnut, rice and corn by dividing cultivated land into three plots.

Along the farm boundary, about 1000 seedlings of mezali (*Cassia siamea*) and about 1000 seedlings of yamanae (*Gmelina arborea*) were planted with the purpose of producing fuelwood, poles, timber and manure. Trees around the farm boundary are planted at a spacing of 15 ft. × 15 ft.. As per the layout design indicated in Figure 4, some user group members

Table 1   Operational schedule for intercropping

No	Activities	Operational season (month)											
		1	2	3	4	5	6	7	8	9	10	11	12
1	Soil preparation		■	■									
2	Seedling collection				■	■							
3	Fertilizer gathering				■	■							
4	Planting fruit & forest trees					■	■	■					
5	Crop planting												
	i. paddy						■						
	ii. groundnut								■	■			
	iii. corn					■	■						
6	Harvesting												
	i. paddy									■			
	ii. groundnut										■	■	
	iii. corn											■	■

also plant forest trees by mixing fruit trees such as jackfruit and mango along the farm boundary. Table 1 gives an indication of the calendar of activities for intercropping.

Community forestry for conservation of natural forests

The user group also manages 15.89 ha of natural forests that are particularly aimed at rehabilitation. According to the field inventory, there are about 31 tree species in natural forest area. To conserve existing natural forests, proper silvicultural treatments such as improvement felling, enrichment planting, weeding, pruning, thinning and fire protection are being carrying out. Table 2 shows the seasonal plan for each activity, workload sharing system and how the various activities are implemented integrating traditional knowledge with modern science through a participatory decision making process. The species preferred for gap planting are teak (*Tectona grandis*), Bamboo (e.g. Wani *Dendrocalmus latiflorus*) and yamane (*Gmelina arborea*) which would be planted in 2014.

Table 2   Management plan for conservation of natural forests

Activities	Season	Specific activities	Work sharing system	Remarks
1. Land allocation for CF management in the reserved forest	August – September	Natural forests are divided individually through users' consensus	Collective decision	Nil
2. Improvement felling	September – October – November (will be conduced every year)	Cutting inferior trees for developing healthy and sound trees; climber cutting and coppicing will be carried out	Individually	For these operations, staff from the Forest Research Institute provided practical training by working together with user group members

(to be continued)

Activities	Season	Specific activities	Work sharing system	Remarks
3. Fire prevention	February – March	Outer fire line is 6 ft. wide while inner line is 3 ft	Collective action	Fire protection will be conducted every year
4. Planting trees (enrichment planting)	May – June – July	Rare, endangered and locally preferably species will be planted	Individually	About 250 tree seedlings would be planted in 2014
5. Weeding		3 times of weeding will be carried out	Individually	
6. Rotation for harvesting	7 years for fuelwood; 25 years for poles; 30 years for posts;	Selected trees will be harvested using coppice selection system	Individually	

Home gardens

Majority of the villagers are Danu who have a tradition of establishing and managing home gardens (Figure 6). Houses are surrounded by several annuals and perennials. Common annual crops are egg-plant, pumpkin, lemon, various kinds of beans, chilli, ladies' finger, bitter guard, mustard and tomato. Perennials (including fruit trees) include banana, papaya, mango, coconut and coffee. In addition to edible crops and fruit trees, they also plant ornamental plants such as rose, jasmine, kiss me quick, temple tree (*Plumeria acutifolia*) and ywet hla (*Musaenda luteola*).

Figure 6 A home garden

For studies on enhancement of home gardens, 55 households (one fourth of total households) were randomly selected and their preferred crops, fruit trees and forest trees were identified through household surveys. Interview results indicated that home garden products partly contributed to subsistence consumption and small amount of income as well. The preferred species so identified include improved varieties of jackfruit, papaya, mango (*Mangifra indica*), sandal wood (*Santalum album*), agar wood (*Aquilaria malaccensis*) and coconut. In 2014, about 500 mango seedlings were distributed to the local communities. Seedlings of other preferred species would be provided during the coming planting season (Box 1).

> **Box 1**
> **Work plan for 2014**
>
> **1. Nurseries for production of seedlings and other planting materials of rare and endangered native tree species**
>
> (1) Field experimentation of site requirements and techniques to prepare and transplant seedlings and other planting materials of locally preferred, rare and endangered native tree species.
>
> (2) Field demonstration of site requirements and techniques to prepare seedlings and planting materials of locally preferred, rare and endangered native tree species.
>
> (3) Completion of the database of locally preferred, rare and endangered native tree species (list of species, site requirements and propagation techniques, based on results of Year I and Year II in this regard.
>
> **2. Techniques for soil improvement in degraded areas for tree planting**
>
> (1) Soil improvement and rehabilitation of degraded forestland.
>
> (2) Field experimentation of techniques to improve soil conditions of degraded areas for tree planting.
>
> (3) Field demonstration of techniques to improve soil conditions of degraded areas for tree planting.
>
> **3. Agroforestry systems, including under-storey cultivation**
>
> (1) Upland agroforestry-based models, including understory cultivation.
>
> (2) Field experimentation of agroforestry models in line with the participatory planning of forest rehabilitation.
>
> (3) Field and on-farm demonstration of agroforestry models, including indigenous practices.

Conclusions

As indicated in the baseline socio-economic assessment of the target village, taung-ya has been a widespread practice to provide subsistence food and income to the local communities. Also it is a culturally deep-rooted practice making it difficult to eliminate it completely. Further, forests provide fuelwood for domestic consumption as also to generate income, especially during the seasons when there is no income from the farms. Sustainable forest rehabilitation will have to take this into account to ensure effective local community participation. Forest rehabilitation activities in the coming years will focus on strengthening the participatory process involving local communities, private sector, local officials, elder people and scientists. Meanwhile, training and extension activities will be conducted to raise awareness on forest resource management, sustainable agricultural practices and income generating activities.

Potential and Prospects of Community Forest Management in Nepal: a Case Study of Panchase Community Forest

Rom Raj Lamichhane[1]

Abstract

The paper provided an overview of ongoing efforts in Nepal in sustainable forest management and biodiversity conservation particularly focusing on local community involvement. Nepal is a pioneer in pursuing community involvement in forest management through local forest user groups. Drawing upon the experience of the Panchase protection forests, the paper outlined the approach adopted for integrating biodiversity conservation, watershed protection and ecotourism with the enhancement of the livelihood of local communities. The paper underscored the importance of equitable sharing of benefits and a pro-poor approach especially in leasing marginal lands to poorer households.

Introduction

Spreading over an area of 147,490 km^2, Nepal lies between China and India extending about 800 km from the East to the West and 160 km from the North to the South. Panchase, the study area, is bounded with 9 Village Development Committees of Kaski, Parbat and Syangja Districts of the western development region. The region has great biological, cultural, and religious diversities and natural beauty as well. It represents an important middle mountain ecological zone which is inadequately represented in the economy's protected area system and is the only corridor linking the Lowland (Chitwan–Nawalparasi) and the Annapurna

---

1. Planning Officer of Ministry of Forests and Soil Conservation

Himalayan Range. The Panchase Protected Forest has an area of 57.76 km^2 and was gazetted in 2012. With the altitude varying from 1,450 m to 2,517 m, the forest is characterized by subtropical and temperate vegetation. The historically famous "Panchase" lake is situated at an altitude of 2,250 m in the area. Having the highest percentage of forests in year 1986 and second highest population percentage as per the 2012 national census, the middle mountain region has witnessed a high rate of deforestation of 2.3 percent between 1978 – 1979 and 1990 – 1991. Taking into consideration its rich cultural and biodiversity values, the government of Nepal has declared the Panchase area as a protected forest under the *National Forest Act 1993*.

Nepal is a pioneer in implementing community forestry, involving local communities in the management of forests through the system of forest user groups. So far, about 1.2 million ha of forest land have been handed over to more than 23,000 Community Forestry User Group (CFUGs). Thirty five percent of the total population has benefited from this program. The community forestry program has been successful in improving the natural regeneration and productivity of the forest. The impacts of community forestry have been subjected to in-depth studies indicating its positive biophysical impacts. Barren and waste lands have been reverted to green forests. Users have security of tenure—access, use, and control—on forest resources. Besides, fulfilling basic needs, many CFUGs have increased their income by way of selling surplus forest products significantly contributing to community development and enhancing household income. This has made community forestry as one of the most successful participatory approaches to natural resource management in Nepal as also globally.

Flora and fauna

With 5 forest types, the Panchase area is very rich in biodiversity. Kharsu (*Quercus semecarpefolia*), phalat (*Quercus* species) and lali Gurans (*Rhododendron arboretum*) are the major species in the upper reaches of the area, whereas chilaune (*Schima walichii*), katus (*Castanopsis indica*), rakchan (*Daphniphyllum himalayense*) and sal (*Shorea robusta*) are found in the lower belt. More than 589 flowering plant species have been recorded in the Panchase forests. Of these, 107 species are medicinal plants, 8 are fibre yielding plants, 23 are natural dye yielding plants and 18 wild species have potential for floriculture (except orchids). In addition, there are 56 species of wild mushrooms and 98 species of ferns found in the area. This region is commonly known as the Kingdom of Wild Orchids. Among the 412 species of orchids reported in Nepal, 113 species including two endemic species (*Panisea panchasenensis* and *Eria pokharensia*) are found in the area with 35 of these having high commercial value.

Wild animals found in the forest area are Asian black bear (*Selenarcto thibetanus*), barking deer (*Muntiacus muntjak*), leopard (*Panthera pardus*), jungle cat (*Felis chaus*), fox (*Vulpus vulpus*), jackal (*Canis aerus*), wolf (*Cania lupus*), monkey (*Macaca mulata*), rabit, mongoose, *etc*. Some 9 species of bats have been recorded from this area. Important bird species found in the area are Nepali kalij, wild cock, koel, red whiskered bulbul, crow, hawk, owl, and sparrow. Demoiselle crane (*Anthropoides virgo*) and parrot (*Psittacula himalayana*) are the important migratory birds found in the area. The lower belt of Panchase hill forest nearby the settlement is being managed by local people as community forests while the upper areas are kept as protected forests.

Uncontrolled grazing, infrastructure development (especially road construction) and other anthropogenic pressures are the major factors contributing to deforestation and degradation in the Panchase area. The extensive use of forest for various purposes has led to forest degradation.

- Thousands of livestock, mainly buffaloes, freely graze the area round the year under an open grazing system.
- Species like *Michelia champaca*, *Prunus* spp. and *Taxus* spp. are on the verge of extinction and the numbers of *Orchid* species are also declining.
- Wildlife population is under threat due to increased poaching.
- Soil erosion and sedimentation of the Phewa Lake is another major problem of the area.

The main aim of the forest conservation is the sustainable management of biodiversity, water resources and ecotourism through participatory management approach ensuring environmental stability and livelihood promotion of local people. The program aims to accomplish the following:

- Community-based sustainable management of flora, fauna, habitat and micro ecosystem; wise use of forest products and conservation of natural beauty of the area,
- Community-based ecotourism and creation of environment-friendly jobs in order to improve livelihood of local people; and
- Participatory watershed management within the Protected Forest, ensuring mutual relationship between up-stream and down-stream by establishing payment for environmental services.

The protection activities in the upland area could considerably reduce soil erosion, sedimentation, and flooding in the downstream. As upland communities rarely receive the benefits of the environmental services that they provide, there is a realization of the need for transfer payments for the environmental services that they provide to improve their livelihoods. It is in this context that efforts are being made to put in place a system of payments for environmental services.

To accomplish the above objectives, four broad programmes—Sustainable Community Forest Management, Biodiversity Conservation, Soil and Watershed Conservation and Local livelihood Enhancement—are being implemented with organizational structures at the District Council and Local Council levels. The keys to the involvement of local communities are (a) an opportunity to share the benefits equally, and (b) a pro-poor approach through leasing marginal lands to the poorer households. Further, an effective participatory monitoring system fully involving all stakeholders and users as also involving officials at the district and regional forest department is in place, in addition to monitoring by expert groups.

Problems, solutions and the way forward

As with all programmes dealing with conservation and development, one has to address a number of challenges during the course of implementation of the programme. Some of the problems and how they are tackled are indicated in the Table 1 below.

Table 1 Problems, Solutions and the way forward

Problems and issues	Solutions and the way forward
Lack of budget will be a hindrance to achieve goals and progress	Manage the budget by Department of Forestry (DoF) as yearly regular budget and cooperate with other helping institutions
Free grazing and traditional rotational grazing leads to deterioration of biodiversity	Provide other income generation options
Lack of legal authority to punish poachers and herb smugglers	DoF and regional directorate to prepare the rules and regulations providing authority to local forest council
Siltation of water, especially the Phewa lake	Payment for ecosystem services needs to be implemented to encourage protection of watersheds
Local communities not taking up responsibility for conservation and government not in a position to protect the areas	Mobilize local council and appoint more employees from the department of forests
Problem from forest fire and rural infrastructure like roads results in increased soil erosion	Provide fire control training to local people and provide proper equipment to them; implement green roads provision
Lack of integrated ecotourism system	Develop integrated ecotourism

Panchase biodiversity management project

A unique aspect of the Panchase area is the lake, which apparently has no springs and is totally fed by precipitation requiring that the catchments are effectively protected. Being located in the saddle of the middle hills, the lake has several cultural, historical and amenity values, requiring coordinated efforts. It is in this context that Government of Nepal is implementing the Panchase Biodiversity Management Project (PBMP) with the support from the UNDP and GEF.

Goals and objectives

The long term goal of the Project is to manage the Panchase forest ecosystem as an integrated, community-based "biodiversity-landscape management project", such that its geographic integrity, biodiversity richness, ecological processes and hydrological functions are maintained, while also ensuring that the requisite natural resources, or their alternatives are equitably distributed, concomitantly diversifying and strengthening local livelihoods in a sustainable manner, and helping alleviate poverty and improve the quality of life for a broad array of local people. The specific objectives of the Project are:

- To develop and strengthen local organizational and institutional capacities for good governance and improved service delivery;
- To increase awareness and adoption of alternative energy sources and other "green" technologies;
- To improve community-based biodiversity management; and
- To develop diversified income-generating activities.

Project area

The Project covers the Panchase Lake and it's adjoining three Village Development Committee (VDCs) of three districts (Arther Dandakharka – Parbat, Bangsing – Syangja and Bhadure Tamangi – Kaski). After the completion of first phase of PBMP and assessing its effectiveness, implementation of the second phase was commenced in January 2008 and the Project extended to cover two additional VDCs in the Parbat District, namely, Ramja Deurali and Chitre.

Implementation

The Panchase Biodiversity Management Project is being implemented by the Machhapuchchhre Development Organization (MDO) with the support of UNDP/Global Environment Facility – Small Grants Programme (UNDP/GEF-SGP).

Assessment of the Project

Protected areas form the core of conservation strategies in all the economies, and they are set aside to maintain the ecosystem processes in an intact manner. An assessment was made on the effectiveness of the Project and a team was established with the following objectives:

- To critically survey the Panchase area and identity the most suitable community-based management modality in consultation with local communities and other stakeholders;
- To clearly define a governance structure for the management of the area;
- To define necessary preconditions for successful declaration of Panchase as a protected area;
- To formulate a comprehensive and high standard management document to effectively manage the Panchase area; and
- To provide all the available literatures for developing the document.

The final report of the study was submitted to the National Trust for Nature Conservation (NTNC) and an interaction workshop for sharing of the study findings was organized. The impacts of the Project are summarised below.

Impacts of the Project

Effects on biodiversity

The vegetation of Panchase forest has been improved considerably in comparison with the earlier situation. This has been accomplished through massive plantation activities undertaken with the participation of local communities and the technical and financial support provided by the Project. Likewise, the Panchase area is the main watershed area of the Phewa Lake. In the last few years deposition of silt in the Phewa Lake has increased and the Project has been helping to control soil erosion in the watershed area. Open grazing has been controlled and so is encroachment. Local people has planted fodder and forest trees in the open land, homeland and other unutilized places, thus reducing the extent of barren land and increasing the greenery.

Effects on climatic change

The main focus is to provide improved cook stoves, enhancing the efficiency in the use of woodfuel thus reducing carbon emissions. The use of improved cook stoves has been picking up rapidly. Another option being popularized is the use of solar tuki, which has been replacing the kerosene lamps thus reducing carbon emission.

Sustainable livelihoods

The Project has carried out a number of activities to promote sustainable livelihoods. It has carried out training to develop human skills, especially to manage NTFPs at the local level. Farmers have started seasonal and off-season vegetables farming, bee keeping, coffee cultivation, *etc*. These activities have generated additional income for the people. Saving and credit program through group formation has provided sustainability to investments and incomes. Proper management and effective implementation of the Project has increased the funds of the group. Locally available resources—natural, human and financial resouces—have been well mobilized and utilized. Forest conservation and improved management has eased their lives, improving availability and access to fodder and fuel sources. Support has also been provided to establish micro enterprises.

9 Degraded Forest Rehabilitation and Sustainable Forest Management in the Philippines

Aurea P. Lachica[1]

Abstract

The Philippines have experienced extensive deforestation and degradation over the last century. The economy lost its forest cover rapidly due to heavy logging, upland migration and agricultural expansion. Consequently, it faces timber shortages and relies heavily on imports to meet a large proportion of its demand. Severe flooding and landslides occurring almost annually are often attributed to deforestation, which has paved the way for a number of government initiatives in reviewing and implementing forest protection works. Massive efforts to rehabilitate/restore degraded forest lands in the economy stem from the economic, social and environmental impacts of deforestation and forest degradation. The paper provided an overview of two major programmes currently under implementation, namely the National Greening Programme and the moratorium on logging in the natural forests and the creation of an anti-illegal logging task force. The National Greening Programme aims to plant 1.5 billion trees in 1.5 million ha through a highly participatory approach coordinating the involvement of all stakeholders.

Introduction

The Philippines was once endowed with rich natural resources, verdant mountains, blue water, clean air and rich flora and fauna. Among these, the forests in the economy are

---

1. Senior Forest Management Specialist of Forest Management Bureau, Department of Environment and Natural Resources (DENR), Philippines

considered as vital on account of the tangible and intangible benefits that they provide. These benefits include timber and timber-based products, non-wood forest products (resins, vines, honey, *etc*.) and more importantly the intangible benefits include recreational and aesthetic values. This has made forests as the centre-piece of the economy's natural resource base. It is in view of this critical importance that the Spanish regime has created the forestry service to conserve, manage and develop the forest resources, developed the forest industry and initiated various forestry research programmes.

The total land area of the Philippines is 30 million ha, which have been classified into (a) alienable and disposable land, and (b) forest land. As of 2012, the extent of area classified as forestlands is 15.05 million ha or 50.2 percent of the land area. In addition, Philippines has some 0.755 million ha (or 2.5 percent of the land areas) of unclassified forestland. The classified forest land of 15.05 million ha has been grouped into the following categories.

- Established forests: 10.056 million ha;
- Established forest reserves: 3.270 million ha;
- National parks, game refuge, bird sanctuaries and wilderness areas: 1.340 million ha;
- Military and naval reservation: 0.126 million ha;
- Civil reservation: 0.166 million ha;
- Fish ponds: 0.091 million ha.

In 1900, out of the 30 million ha total land area of the economy, the forest cover was 70 percent or 21 million ha. It was reduced to 60 percent equivalent to 18 million ha and 50 percent or 15 million ha in 1920's and 1950's, respectively. In 1960's, the remaining forest cover was only 10.2 million ha or 34 percent and which was further reduced to 7.4 million ha or 24.7 percent in 1980. According to estimates, forest cover at the beginning of 1999 was about 5.5 million ha or 18 percent of the economy's land area, but in the 2003 statistics, a 1.8 million ha increase was recorded. This is mainly attributed to the reforestation efforts by the government and private sectors through the development of industrial tree plantations and the management of natural forest areas within the Community-based Forest Management, *Socialized Integrated Forest Management Agreement* as well as afforestation of private lands through the *Private Land Forest Development Agreement* and other reforestation initiatives. Yet based on the 2010 satellite imageries, the total forest cover of the Philippines recorded a decline to 6.840 million ha in 2010 from 7.168 million ha in 2003. This means that between years 2003 and 2010 the economy lost around 328,682 ha of forest area, amounting to an annual decline of 46,954 ha.

The decline in the Philippines forest cover is attributed to the continuous increase of the population and the pressure that it exerts on land for farming, urbanization and use of land for other purposes including infrastructure development. Over exploitation of timber and other non-wood forest products, ineffective implementation of forestry sector policies and the weak institutional arrangements are the primary causes of deforestation and degradation.

In response to this alarming situation, the government has implemented various measures to address the problem of deforestation and the worsening condition of the forests. The massive rehabilitation/restoration of degraded forest through reforestation program, assisted natural regeneration, plantation establishment, agroforestry and the like are the priority task of the environment sector. These activities were also incorporated in the basic *Forestry Laws* issued decades ago. These policies are as follows.

• The *Presidential Decree (PD) 705* known as the *Revised Forestry Code of the Philippines* adopted the multiple use of forestland oriented to the development of the economy as well as the protection, development and rehabilitation of forestland so as to ensure improvement in forest productivity.

• The *1987 Philippine Constitution* stipulates that "the exploration, development and utilization of natural resources shall be under the full control and supervision of the State, thus it is the obligation of the State to protect the remaining forest cover not only to prevent flash floods , but also to preserve biodiversity, protect threatened habitats and sanctuaries of endangered and rare species, and allow natural regeneration of residual forests and development of plantation forests".

• The Executive Order No. 263 known as the "Community-Based Forest Management Program" serve as the national strategy to achieve sustainable forest management in the economy promoting forest conservation and development activities through collaborative undertaking of all forestry stakeholders ensuring that local cultures and rights of indigenous people to their ancestral lands are respected and recognized.

• The Executive Order No. 318 (2004) aimed to promote sustainable forest management in the Philippines includes (a) delineation, classification and demarcation of state forest lands; (b) holistic, sustainable and integrated development of forest resources; (c) community-based forest conservation and development; (d) proper valuation and pricing of forestry resources and financing SFM; (e) incentives for enhancing private investment, economic contribution and global competitiveness of forest-based industries.

In order to accelerate forest recovery in the economy, in 2011 President Benigno S. Aquino III has issued Executive Order No. 23 *Declaring a Moratorium on the Cutting and Harvesting of Timber in the Natural and Residual Forest and Creating the Anti-illegal Logging Task Force* and Executive Order No. 26 *Declaring the Implementation of the National Greening*

Programme (NGP). The enforcement of forestry laws and regulations has been strengthened for the smooth implementation of any development and management endeavour.

Approaches/ policies/ programs toward rehabilitation of degraded forests

National Greening Programme (Executive Order No. 26)

The National Greening Programme (NGP) launched on the basis of Executive Order No. 26 in February 2011 forms the largest reforestation program in the Philippine history. This is part of the *Social Contract of the President* aiming to plant 1.5 billion trees in 1.5 million ha during 2011 to 2016, in public domain lands including forestlands, mangrove and protected areas, ancestral domains, civil and military reservation and urban areas. In terms of magnitude, this is more than what has been planted in the past 50 years. The NGP intends to address poverty reduction by providing alternative livelihood activities to marginalized upland and lowland households through their involvement in seedling production and the care and maintenance of newly-planted trees. Apart from being a reforestation initiative, the NGP is also seen as a climate change mitigation strategy as it seeks to enhance the economy's forest stock for absorbing carbon dioxide. It also intends to address food security and biodiversity conservation. Table 1 gives the annual planting targets in number of seedlings and the area to be covered under the National Greening Programme.

During 2011, the first year of the Programme, about 89.62 million seedlings, which is equivalent to some 128,558 ha of open and denuded forestland, have been planted. This is about 17 percent above the target for 2011. During the last three years, an equivalent of 683,481 ha has been planted (Table 2). The target for planting in 2014 is 381.5 million trees. Aside from the benefits from additional forests, the Programme has so far employed more than 1,182,000 persons from upland and rural communities. The Programme has produced high-value food and cash crops such as fruit trees, coffee, cacao, rubber and others.

Table 1 Annual target under NGP

Year	Target	
	Area (ha)	Seedlings (in millions)
2011	100,000	100
2012	200,000	200
2013	300,000	300
2014	300,000	300
2015	300,000	300
2016	300,000	300
Total	1,500,000	1,500

The National Greening Programme is largely a community-based programme and

Table 2 NGP—Targets and accomplishments during 2011 – 2013

Year	Target area (ha)	Area planted (ha)	Accomplishment in percentage (%)	Number of seedlings planted	Number of jobs generated
2011	100,000	128,558	129	89,624,121	335,078
2012	200,000	221,763	111	125,596,730	380,696
2013	300,000	333,160	111	182,548,862	466,990
Total	600,000	683,481	114	397,769,713	1,182,764

reforestation activities are contracted to communities. Seedlings are grown in a network of 22 clonal nurseries and nurseries in the 26 state universities and colleges (SUCs). Efforts are underway for the establishment of mechanized nurseries to improve the quality of planting material, and thus to reduce the cost of seedling production. Those that are not grown in the Forest Management Bureau (FMB) nurseries (e.g. coffee, cacao, rubber) are procured through competitive bidding, in accordance with the procurement law. Private sectors, civil society organizations and local government units also participate in these efforts.

All sites where planting is done are geo-tagged and pictures of progressive changes are taken with the global positioning system (GPS) reading and date and time. It is also proposed to use unmanned aerial vehicles to monitor and validate the reforestation efforts. The progress reports submitted by field personnel are all notarized by lawyers and every effort is made to ensure good governance guaranteeing that funds allocated for the purpose are effectively and efficiently used for reforestation.

It is estimated that about half of the total budget of the National Greening Programme (or about P 30 billion—approximately US$ 682 million) will go directly to the communities as wages through job creation, contributing to inclusive growth in the rural upland areas.

Nationwide Logging Moratorium on Natural and Residual Forest (Executive Order No. 23)

While investing in the rehabilitation of denuded and degraded forest land under the National Greening Programme, Philippines is also stepping up efforts to better protect the remaining forested areas to accomplish the forest cover target of 30 percent of the geographical area by 2016 envisaged under the *Philippine Development Plan*. In order to stop further forest depletion, the government has imposed a nation-wide moratorium on logging in all natural forests in 2011 through the issuance of Executive Order (E.O.) No. 23 *Declaring a Moratorium on the Cutting and Harvesting of Timber in the Natural and Residual Forest and Creating the Anti-illegal Logging Task Force*. The Executive Order 23 together with intensified enforcement has led to the confiscation of 25.5 million board feet of illegally cut and processed forest products. About 1,233 cases have been filed in the courts and hitherto

186 persons have been convicted. This has led to a reduction in the number of illegal logging hotspots by 84%, from 197 municipalities in 51 provinces to just 31 municipalities in 12 provinces in a year.

To protect and expand the forest cover, a proposal costing of about P 906.27 million has been formulated especially to strengthen the protection and law enforcement capabilities based on the updated *Forest Protection Plan* of each region. A menu of options and specific strategies based on the specific geographical conditions of the provinces has been drawn up. This Menu of Options comprises of a list of strategies and a set of activities for every strategy, providing immediate and long-term actions to be pursued to accomplish the goals of the National Forest Protection Programme.

There are the following eight components of the Menu of Options:

- Provision of full logistic material support essential for forestry law enforcement;
- Improvement of infra-structure, provision of institutional support in investigation, filing of information and/or criminal complaints and prosecution of forestry cases;
- Active collaboration and involvement of forest communities and other stakeholders in forest protection and law enforcement undertakings;
- Undertaking of capacity building for DENR field personnel to enhance their skills and competence for effective protection of forest and plantations for biodiversity conservation;
- Development of well-planned information, education and communications (IEC) campaign region-wide;
- Consistent apprehension and mandatory administrative adjudication and confiscation of undocumented forest products including conveyances and implements;
- Effective forest fire, pest and diseases management measures; and
- Institute forest certification and timber legality and assurance systems and other reforms.

Implementation of the above Menu of Options is expected to increase the forest cover, improve forest ecosystems, mitigate climate change, and enhance the flow of environmental services, helping to accomplish the objective of sustainable forest management in compliance with the Philippine Criteria and Indicator System.

Lessons learned

The Programme is expected to be successful on account of the strong support from government. Philippines has the required technical capacity backed by long experience in the technical and organizational aspects of reforestation and rehabilitation. Further, the National Greening Programme has broader objectives of poverty alleviation, improvement of food security, conservation of biodiversity and climate change mitigation and adaptation. There is therefore considerable optimism as regards the attainment of a forest coverage target of 30 percent by 2016.

The way forward

Although a lot has been accomplished as regards reforestation and forest protection, more needs to be done. This includes a clear indication of where the protection and production forests are to be located. As an initial step, there is a need to effectively engage local government units in the sustainable management of the forests through investing in forest land use planning and later integrating this with the overall land use plan of every municipality.

Summaries

The Philippine government, in particular the DENR is making concerted efforts to rehabilitate the open and denuded forest areas through several initiatives. The National Greening Programme (NGP) implemented in accordance with Executive Order No. 26 is a priority programme of the government. During the 6 years of the NGP, some 1.5 billion trees will be planted. The NGP intends to pursue sustainable development for poverty reduction, food security, biodiversity conservation and climate change mitigation and adaptation. This consolidates and harmonizes all the greening efforts of the government. Further, it is not just planting of trees as it includes more employment and income-generating crops like coffee, cacao, rubber, bamboo, rattan and other fruit trees.

Implementation of Executive Order No. 23, *Declaring a Moratorium on the Cutting and Harvesting of Timber in the Natural and Residual Forest and Creating the Anti-illegal Logging Task Force* is another important component of the forest rehabilitation and protection efforts in the economy. Through this, the forest protection activities will be strengthened through effective enforcement of forestry laws, rules and regulations. As in the case of NGP, the implementation of Executive Order 23 visualizes the active participation/ involvement of various stakeholders and continued capacity building of DENR field personnel.

It is expected that by the end of the two programmes forest loss in the economy would be reversed and degraded forests would be brought back to increased productivity. Forest coverage is expected to increase from 24 percent to 30 percent of the land area. However, achieving optimum results is dependent on strong collaboration between different stakeholders including various government agencies, private sectors, local government units (LGUs), other governments and non-government organizations.

References

Forest Management Bureau. December 2013. Department of Environment and Natural Resources. Philippine Forests Facts and Figures.

DENR-FMB. December 2013. Paper on Five Year National Forest Protection Program Menu of Options for Effective and efficient Forest Protection and law Enforcement of the DENR.

Forest Management Bureau DENR. 2014. FMB Technical Bulletin No. 6: Engagement of the Local Government Units in the National Greening Program.

Forest Management Bureau. October 2013. FMB Technical Bulletin No. 4. Specifications/Standards for NGP's High Value Crops Cacao, Coffee and Rubber.

DENR-DILG Joint Memorandum Circular No. 2013-03: Guidelines in the Establishment and Implementation of Barangay Forest Program in Support of the National Greening Program. 30 May 2013.

Executive Order No. 23: Declaring a Moratorium on the Cutting and Harvesting of Timber in the Natural and Residual Forest and Creating the Anti-illegal Logging Task Force. February 1, 2011.

Executive Order No. 26: Declare the Implementation of National Greening Program. February 24, 2011.

Executive Order No. 318: Promoting Sustainable Forest Management in the Philippines.

Executive Order No. 263: Adopting Community Based Forest Management as a National Strategy to Ensure the Sustainable Development of the Economy's Forestlands Resources and Providing Mechanism for Its Implementation.

Presidential Degree No. 705: Revise Forestry Code of the Philippines. 1975.

Forest Management Bureau (FMB), Department of Environment and Natural Resources (DENR). National Report to the Fifth Session of the United Nations Forum on Forest (UNFF).

10 Rehabilitation and Restoration of Forests in Sri Lanka

W.T.B. Dissanayake[1]

Abstract

As in the case of most other economies in the Asia-Pacific, the area under natural forests in Sri Lanka has dwindled rapidly and in 2010, it accounted for less than 24 percent of the economy's total land area. Factors contributing to deforestation and forest degradation in Sri Lanka are many and complex. Poverty associated with the landlessness and the unfavourable land-tenure systems are the main causes. Other causes are large-scale agricultural and human settlement projects, shifting cultivation, and over-exploitation of forest products. Natural causes such as cyclones, tsunami, droughts, forest fires, wild elephant damages, *etc.* also lead to degradation and forest cover decline.

The Forest Department has undertaken forest rehabilitation and restoration programs since 1887 and the overall outcome has been mixed. Introduction of social forestry in the late 1980s has been an important milestone in this regard. A number of projects and programmes consisting of enrichment planting, development of buffer zones, agroforestry woodlot establishment, *etc.* are being carried out nationally with the involvement of local people and communities.

This paper analysed the different approaches adopted in Sri Lanka for the rehabilitation and restoration of degraded forests, outlining the issues and challenges faced in the process and emphasizing the need to develop partnership with local people, communities, NGOs, and the local private sector.

---

1. Conservator of Forests, Planning and Monitoring of Forest Department, Sri Lanka

Introduction

Sri Lanka was once a heavily forested island, but during the last century the extent of closed natural forests has dwindled rapidly to less than 24 percent of the total land area. Historically, much of the loss was attributed to the creation of plantations of tea, rubber, and coconut and other crops during colonial times. More recently, settlement schemes to provide livelihood for the landless poor, slash and burn agriculture, unplanned encroachment, expansion of human settlements due to population growth, illegal logging, uncontrolled harvesting of fuelwood, and forest fires have contributed to the decline of island's natural forest cover.

Sri Lanka is one of the smallest but biologically most diverse as well as highly populated economies in Asia. Consequently, it forms part of a biodiversity hotspot of global and national importance. Its varied topography and tropical island conditions have given rise to very high levels of biological diversity and endemism, particularly in the natural forests. The ecological value of these forests is recognized even internationally. In addition, the forestry sector in Sri Lanka is closely linked to the agriculture and energy sectors. The main source of electricity is hydropower, which very much depends on a regular supply of clean water. Similarly, the performance of agriculture sector depends on regulated supply of irrigation water and maintenance of soil fertility, which are important functions of forests.

Sri Lanka has a strong tradition in forest conservation. Its system of protected areas is one of the most extensive in Asia and covers more than 28% of the total land area. About 1,846,000 ha of the total land area is reserved and administered by either the Forest Department (FD) or the Department of Wild Life Conservation (DWLC).

As per the national income statistics, forestry sector's contribution to the gross domestic product (GDP) in 2013 was just 0.4 percent. However, the true contribution to the economy is much greater considering that a significant share of the products and services provided by forests goes unaccounted especially on account of the lack of statistics. A sizeable share of production of sawn wood, fuelwood, and non-wood forest products are not accounted at all. There are also challenges in accounting the environmental services provided by forests. A conservative estimate by the Forestry Sector Master Plan of 1996 puts the forestry sector's contribution to the GDP as 6 percent.

National forest policy

The 1995 national forest policy of Sri Lanka acknowledges the importance of protecting the remaining natural forests for posterity in view of their importance in the conservation of biodiversity, soil and water resources. The policy emphasizes the importance of retaining the present natural forest cover, and of increasing the overall tree cover. It also recognizes the crucial role of home gardens and other agroforestry systems, and of trees on other non-forest land in supplying timber, bio-energy, and non-wood forest products. Further, recognizing that the state alone is unable to protect and manage the forests effectively, the policy emphasizes the need for people's participation, especially through partnerships with local people, communities, NGOs and the local private sector.

Forest degradation

A complex web of factors, many of which are outside the forest sector, contribute to deforestation and forest degradation in Sri Lanka. Obviously, the root cause is socio-economic in nature with poverty associated with landlessness and a poor land tenure system being the primary driver. Other causes of deforestation are large agricultural and human settlement projects, such as Mahaweli project, along with its reservoir and hydropower projects, chena (shifting) cultivation, excessive harvesting of forest products, and the conversion of natural forests to plantation and arable land.

In a predominantly agricultural economy like Sri Lanka, a strong link exists between population growth and deforestation and forest degradation, especially in view of the increasing demand for various products. Increase in agricultural production is accomplished mainly by expanding the area under cultivation. Since most of the other cultivable land has been already in use, further expansion of cultivation has been realized through the clearance of natural forests. Consequently, forest area per capita in Sri Lanka has declined from about 1.3 ha in 1900 to less than 0.1 ha in year 2010. As the population keeps growing, the remaining natural forests are under increased pressure.

Forest resource depletion is also closely linked to the demand for forest products such as timber, non-wood forest products and fuelwood. Population increase combined with economic growth has resulted in higher demand for these products.

Fire is another major cause of deforestation and forest degradation, especially in the dry and intermediate zones of Sri Lanka. Large tracts of natural as well as planted forests are subjected to fire during the dry season of the year, gradually reducing their quality and ultimately turning them into grasslands. In addition, rare occurrence of extreme weather conditions such as cyclones is also causing damage to forest lands. The tsunami in 2004 resulted in the large-scale destruction of coastal forest ecosystems, especially in the eastern and southern parts of the economy.

In addition to the above, policy and institutional problems also contribute to deforestation. Inappropriate land-use is primarily due to insecure tenure and the absence of an explicit land-use policy clearly specifying the development objectives and associated legislation.

Forest restoration

The progressive decline in the forest resources combined with population growth has led to an imbalance in the demand and supply of timber and fuelwood. Consequently, in late 1970s and early 1980s, there has been a widespread recognition of a need for stepping up reforestation and forest conservation efforts in both government and donor circles.

The forest rehabilitation and restoration programmes implemented in Sri Lanka can be categorized firstly by way of their silvicultural approaches and secondly based on their implementation arrangements/mechanisms.

Silvicultural approaches

A complete list of silvicultural options would require an in-depth analysis of the theory and practical variations of different silvicultural systems. A simplified summary of the options, concentrating on the regeneration method, from less intensive to more intensive, would be as follows.

- Assisted natural regeneration: This is the most recent trend in the restoration of degraded forests especially in the dry zone of the economy where the degraded forests are mainly regenerating chena (shifting cultivation) lands and only a few large trees per hectare exists. The main thrust is to protect the regeneration from fire, followed by tending and the application of fertilizers to enhance the rate of growth of seedlings and saplings.

- Enrichment planting with indigenous or exotic species: This option is implemented in the more degraded forests where indigenous trees still occur and have to be preserved, but where the pressure on land is so intense requiring higher sustainable yields than what could be attained through natural process. Species are selected taking into account their suitability to the ecological conditions and the nature of demand to be fulfilled.

- Forest plantations: Conversion to forest plantations is being adopted only when the indigenous forest has deteriorated beyond recovery. This is done after giving due consideration to the local socio-economic and ecological conditions. Tree species are selected according to the desired end uses, taking into account the sustained production potential of the site. Large extent of *Eucalyptus grandis* and *Eucalyptus microchoris* in the wet zone hill economy and teak, *Eucalyptus camaldulensis*, *Eucalyptus teriticornis* and *Acacia auriculiformis* in the dry and intermediate zones have been planted under this system.

- Agro-forestry: This system is the most common and most favored system whenever the local people are involved in reforestation of abandoned chena lands. Trees and field crops are grown together for about 3 – 5 years, but priority is given to the trees. Among the different tree species grown under agroforestry, teak has been particularly successful in the dry zone.

Implementation arrangements

Following are the main implementation arrangements/mechanisms being practiced in Sri Lanka for the rehabilitation of degraded forests.

Block Reforestation

Block reforestation is defined as the reforestation of medium to large size tracks of state land by the Forest Department, either directly (employing labourers directly) or through contracting the task to established contractors. This is often considered as "classical forestry", i.e. the normal work of forest departments. It can create temporary employment in nearby areas, serve the needs of local people, and help protect the environment. Although it is not often done on a participatory basis, it could be planned in cooperation with local people. The underlying motive is usually either conservation (protection of catchments areas or in coastal areas, stabilization of sand dunes) or production, or a combination of both.

Block planting by local level organizations

This is a form of forestry in which land remains in the hands of the state, but reforestation/afforestation is done by a local organization in cooperation with the Forest Department. The motivation of this local-level organization may be environmental, in terms of protection of hill slopes for the benefit of down-stream cultivators or it may be simply financial—providing employment for its members. The local-level agency operates as a contractor, but because it is not strictly in the sphere of the private market, and because it offers opportunities for participation, it is considered to be qualitatively different from a commercial contractor. (Skutch, 1990)

Social forestry approaches

Several different approaches—more than fifteen—have been pursued by the Forest Department in implementing social forestry. They range from those purely focused on conservation (protection of soil and water resources) to those aimed at employment generation and income enhancement and those geared to purely production (economic growth) objectives.

(1) Co-operative reforestation/village reforestation/taung-ya

The Co-operative Reforestation Programme (which is recently named as village reforestation) in Sri Lanka is a modified version of Taung-ya practised in Myanmar. Farmers who practice shifting cultivation are allocated 1 – 2 ha of state lands for a period of 4 years under lease agreements. The farmers are responsible for clearing the land and planting the seedlings. The seedlings, primarily teak (*Tectona grandis*), are supplied by the Forest Department. Cultivators are permitted to grow agricultural crops between the trees during the lease period. They are also paid a cash incentive contingent upon satisfactory performance. This programme was very successful during 1980s when degraded forestlands were available for raising teak. Nearly 12,000 ha of teak plantations have been established under this programme. (Lai and Ariyadasa, 1996)

(2) Farmers' woodlots

The Farmers' Woodlot Programme was one of the major components of the first assisted by Asian Development Bank (ADB) community forestry project assisted by in the central part of the economy during 1980s. It was further enhanced in the subsequent projects and at the moment operated successfully mainly in the dry zone.

Blocks of government land (varying in extent from 0.2 to 1.0 ha, depending on the agro-ecological zone) are allocated to farmers for the purpose of tree growing on a long-term lease of 25 years. In establishing woodlots, the application of agroforestry practices is encouraged as experience has shown that success has been achieved with farmers' woodlots on sites where the soil conditions are satisfactory to support cultivation. Several forest and fruit tree species are offered to participants free of cost. Although a notional spacing of 5m × 2m has been prescribed, it can be adjusted according to the intercropping practices and species chosen. Application of soil and water conservation measures is also encouraged by planting hedges, primarily with *Gliricidia sepium*. Trees planted become the property of the farmers in the earlier projects. However, in the recent projects 20 percent of the final harvest goes to the government.

As the beneficiaries of this programme are resource-poor farmers, various incentives are provided to compensate for the opportunity cost of the time and the effort required to establish woodlots and to undertake the associated soil and water conservation measures. They include food or cash for work, based upon the work norms of the Forest Department. In addition, certain amounts of fertilizer for tree seedlings are also provided. Farmers are also given training if required.

Private sector leasehold reforestation

The programme "private sector participation in reforestation" was launched by the Forest Department in 1995 to mobilise private sector involvement in tree growing in the dry zone lands. Applications were invited from the private sector for the leasing of 10,800 ha of abandoned state lands in five districts. Although a total of 3,572 applications were received, eventually only 715 applicants had sent project proposals in response to the request made by the Department. Based on assessment of the capability (including experience and financial capability), 53 applicants were selected for this programme in April 2000, allocating an area of 507 ha (Zoysa et al., 2002).

Under the leasehold reforestation programme, land is leased to the private sector for a period of 30 years based on an agreement entered into between the forest department and the lessee, stipulating the conditions of the lease. The lessee should primarily engage in reforestation and inter-cultivation of agricultural fruit crops. Animal husbandry is also allowed where it is appropriate. Technical advice required for the establishment and management of plantations is provided by the Forest Department. The lessee should pay an annual rent to the government based on the total value of the leased land.

The performance of this programme has been rather mixed, depending on the interest of the lease holders. About 50 percent of the lessees have commenced their work within the first five years and continued according to the agreement. Some very enthusiastic leaseholders have achieved great success by raising well-managed high productivity plantations. However, there are others who have failed to comply with the conditions stipulated by the Forest Department resulting in the cancellation of their agreements.

Forest plantations as an investment scheme

Establishment of commercial forest plantations by private sector mobilizing investment from the general public is a recent development in Sri Lanka. There are more than 5 private companies operating in the dry zone and intermediate zone using teak and mahogany respectively as their main tree species. This programme is implemented in medium to large scale private lands which are degraded and abandoned. People are invited to invest on a piece of land block where the company will hold the responsibility of planting and managing trees until they become harvestable. Intercropping of fruit species is also done depending on the fertility of the land. The harvest of the intercrop as well the trees will be a property of the investor.

The companies were successful in involving large number of investors mainly from the urban upper middle class through attractive advertisements. However, due to financial mismanagement and exaggerated claims about profitability, performance of many such schemes has been poor. There is increasing concern about misuse of such schemes resulting in significant financial loss to investors.

Non- governmental organizations

The role of NGOs in forest rehabilitation has been mainly indirect. They have helped strengthen rural organizations and build up the confidence of the rural people in finding independent solutions to their problems. However, most NGOs involved in forestry conduct their activities in an *ad hoc* manner. The support of small-scale tree nurseries or community tree-planting initiatives tends to arise as a part of a general programme of rural development in a small geographical area.

Issues and challenges

The emerging picture of the forestry situation in Sri Lanka is not bright. The continuation of current trends will not alleviate the pressure on the forest resources; on the contrary, human pressure on forests is expected to increase. The expanding population base and economic growth will increase the demand for round wood and poles from about 2 million m^3 in 1995 to 2.7 million m^3 in 2020. During the same period, the need for biomass energy will also increase from 9.3 million tonnes to 9.7 million tonnes. At the same time the closed canopy natural forest cover is projected to decline from 22 percent in 1998 to 17 percent by 2020 (Forestry Sector Development Division, 1996).

The forests in Sri Lanka are under immense pressure from diverse users. Industry wants more logs for production of sawn wood and other products, rural people need fuelwood to meet their energy requirements and the growing population requires land for agriculture. At the same time, those concerned about the environmental values of forests want to "preserve" the remaining natural forests intact. Further, international conventions such as the United Nations Framework Convention on Climate Change (UNFCCC) require an increase in tree cover for carbon sequestration. Forestry professionals and other government officials are facing a multitude of problems and responsibilities, with very limited resources to address the problems adequately. Invariably forests have to cater to the conflicting demands of conserving biodiversity, protecting watersheds, providing land for the landless, and supplying a multitude of forestry products to the increasing population.

Reconciling the different, often conflicting demands placed on forests and resolving the conflict between short-term and long-term benefits and costs is not an easy task. However unless corrective action is taken now, future generations are likely to have a worse environment to live in. The capacity of the forests to provide various amenities would be reduced, in many cases irreversibly. Because of the complexity of problems, the time needed for changing people's values and perceptions, and the long-term nature of the forestry, it is apparent that all the problems cannot be solved quickly. The planning horizon has to be long, at least 20 – 25 years. During this time, it should be possible to reach a state that reflects the expectations set for the forestry sector. First, new policies have to be introduced and old ones modified or abolished, strategies have to be formulated, legislative and administrative reforms need to be carried out, and action needs to be pursued to implement the approved policies.

The state has lost some of its control over much of the formerly forested land. In many localities, people have become *de facto* forest managers. However, these people are not protecting the natural forests, as they do not have any incentive to do so. Farmers and businessmen use the forests to fulfil their own short-term demands. The deforested land is not being put back to forestry but being used by the local people for farming, grazing and other uses. On the other hand, it is clear that without people's involvement the negative forestry trends cannot be reversed. The importance of people's involvement and the resources they represent is demonstrated by the fact that most of the forest products needed by the economy have been already produced by the rural people, mainly in the home gardens and the plantations.

Participatory forestry represents an important change in forestry thinking, but it will have to evolve towards people's forestry, which will be people-driven, people-centered, and based on bottom-up planning and decision making. People's participation is important, but in the long term, foresters should also become participants and support the people in their forestry activities. The state's main task will be to develop an environment which will enable the rural people to manage the forests on a sustainable basis.

Partnership with the industrial sector has been regarded less favorably by the society. This form of partnership has been exploitative and inequitable since most of the benefits have gone to a favored few. Very often, the private forest plantations benefit the rich companies or individuals outside the community. They often fail to deliver the promised social and environmental benefits. This may actively harm the poor village communities who are dependent on forest resources for their livelihood. (Foley and Barnard, 1985). The community often views the forest as an inexhaustible resource. Eventually, failure in communication and differences in the socio-cultural background become an obstacle, private forest growers may not adequately appreciate local ideas of land tenure and the productive capability of local inhabitants (Evans, 1982).

There is, nevertheless, a need to develop a partnership with the national industrial sector, since industries provide the engine for development. Compared with all other plantation crops and export crops, where many incentive and subsidy schemes are available, there are not any incentive schemes, such as credit facilities or insurance schemes for the forestry sector in Sri Lanka.

Many NGOs involved in the environmental sector in Sri Lanka have limited social and silvicultural skills required to support participatory forestry programmes. Most often, they reflect the views of the elite and middle classes of Sri Lanka, as well as various external development organizations. However, there is a need to involve them actively in organizing and mobilizing local people and communities to become more active in forest rehabilitation activities.

The way forward

Effective forms of partnership are needed, such as joint forest management and leasehold forestry. It is increasingly being accepted that the state alone cannot manage and protect the forest resources. It needs the active participation of the rural people, communities, NGOs, rural industries and other local private sector groups.

The state

The state must continue to be the highest authority in the forestry sector, which is responsible for forest policy and legislation, policy coordination, law enforcement, the management of the system of protected areas, the issuing of licenses and collection of revenue, and for supporting the other forestry partners and coordinating their development activities.

Local resource users as resource managers

Experience from Sri Lanka and other economies shows that farmers who have security of tenure are ready to make long-term investments. Where security of land ownership or control is created and the rights to harvest and sell trees are secure, farmers plant more trees. Also, people will help protect and manage state forests when they have incentive to do so. This can happen only if it is accepted that the local rural people and communities are "stakeholders" and that they should be involved in forest management in partnership with the state.

The role of the non-state sectors in forest plantation development

Following the government's economic policy, ways of involving local farmers, estate sector, local organized communities, and other private sector groups in forest plantation establishment and management, for example through leasehold forestry, must be developed and promoted. It must be recognized that these changes in the forest plantation sub-sector cannot be effected until a supporting policy, legislation, and a support system such as extension and access to financing are in place. Environmental safeguards also must be introduced.

To encourage the private sector to grow trees in abandoned barren lands in the dry region, provision of commercial incentives and tenure security would be an important requirement. Other incentives should include extension services, especially access to technology and knowhow, credit and infrastructure improvements. These support measures need to be continued until the programme becomes self-sustaining.

The role of non-governmental organizations

In the future, NGO's and various community-based organizations will have to play a greater role in organizing and mobilizing local people and communities to become more active in the forestry sector. NGOs will have to be strengthened and educated to carry out their functions successfully. It is important to try to institutionalize and formalize their involvement in at least some forestry development activities, such as forestry extension, policy formulation and possibly sectoral planning.

References

Evans, J. 1982. Plantation forestry in the tropics. In: Oxford Science Publications. Oxford, UK: Clarendon Press, 187.

Foley G, and Barnard G. 1985. Farm and community forestry, network paper: ODI social forestry network no. 1b. London, 85.

Forestry Sector Development Division (FSDD). 1996. Sri Lanka forestry sector master plan. Colombo, Sri Lanka: Ministry of Agriculture, Lands and Forestry, 380.

Lai CK, and Ariyadasa KP. 1996. Sri Lanka profile. In: Asia-Pacific Agroforestry Profiles. Bogor, Indonesia: Asia-Pacific Agroforestry Network.

Skutch MM. 1990. Social Forestry in Integrated Rural Development Planning. Bangkok, Thailand: Sri Lanka (RWEDP, Field Document no. 24) FAO. 30.

Zoysa M, de Ariyadasa KP Silva, YYK de. 2002. Private forest plantation development in Sri Lanka: issues and challenges. In: Paper presented at the IUFRO Science/Policy Interface Task Force Regional Meeting. Chennai, India. 11.

11
Development of an Integrated Forest Management in Thailand

Utharat Pupaiboon[1]

Abstract

Until the early 1960s, Thailand was very rich in forests with more than half of the nation's territory being covered by forests, supporting the livelihood of people. However, Thailand experienced rapid deforestation during the post-1960 period bringing down the extent of forests to less than 30 percent of the land area by the end of 20th century. Considerable effort has been made by the government to reverse the process, especially thorough reforestation and afforestation. The Royal Forest Department (RFD) has undertaken a number of research projects to develop various techniques in support of ecosystem rehabilitation including improvement of plantation techniques. This paper provided an overview of the various issues relating to changes in forest resources in Thailand, history of deforestation, development of integrated technologies in support of sustainable forest management and issues like tenure, ownership and the overall framework of forest protection strategies.

Introduction

Thailand has a land area of 51.3 million ha and the economy has predominantly been agricultural with 70 percent of the population earning their living in agricultural or related enterprises. Forests and forestlands are the property of the state and under the responsibility and management of the Royal Forest Department (RFD). Section 4 under the *Forest Act*

---

1. Forestry Technical Officer of Royal Forest Department, Thailand

(1941), defines "forest" as land that has not been taken up or acquired by any other means according to any land law. Rapid population growth and economic development have put substantial pressure on Thailand's forest resources.

The total forest area in Thailand by 2006 was estimated at 15.9 million ha or about 30.99 percent of the land area. Forest cover estimates range from 13.0 million ha to 14.8 million ha. Forests in Thailand are classified as: (a) evergreen forests with three sub-types—tropical rainforests, semi-evergreen forests and hill evergreen forests (43 percent of the forest area), dominated by species of the genera *Dipterocarpus*, *Hopea*, *Shorea*, *Lagerstroemia*, *Diospyros*, *Terminalia*, and *Artocarpus*; (b) mixed deciduous forests (22 percent), the dominant species being *Tectona grandis* (teak), *Xylia kerri*i, *Pterocarpus macrocarpus*, *Dalbergia* spp. and *Afzelia xylocarpa*; (c) pine forests (2 percent), mainly of *Pinus merkusii*; (d) mangrove and coastal forests (2 percent), the main mangrove genera being *Rhizophora*, *Avicennia* and *Bruguiera* and the main beach genera *Diospyros*, *Lagerstroemia* and *Casuarina*; and (e) dry dipterocarp forests (31 percent) (FAO, 2005). A network of parks and reserves extends over an area of more than 10 percent of the land area. By 1999, 56 percent of the forest area has been declared as national conservation forests.

It is difficult to define either "afforestation" or "plantation forests" precisely. In particular, it is often not easy to distinguish between afforestation and either rehabilitation of degraded forest ecosystems or enrichment planting, or between plantation forests and various forms of trees on farms. The definition proposed by FAO to the 1967 World Symposium on Man-made Forests and Their Industrial Importance, forms the basis of subsequent official estimates.

Deforestation levels and forest resource depletion in Thailand have been relatively low since the introduction of a complete logging ban in January 1989. Income from the forestry sector has declined to 0.15 percent of real GDP from an average of 0.20 percent in the five previous years. It appears to have had a minimal downstream effect on the overall economic performance (Sumantakul V. and Sangkul S., 2000).

The growth in wood-based industries remains strong, supported by the import of logs and sawn wood from neighboring economies. Further, harvesting of old rubber trees (for replanting) covering an area of 320,000 ha is expected to yield some 20 million m^3 of wood; and this forms an important source of wood supply to the wood-based industries, now and in the future. Because of difficulties with raw material supply to the plywood industry, the demands for panel products, such as particle board and fiber board, have increased relative to the decline in the plywood industry (Sumantakul V. and Sangkul S., 2000).

Regarding the regional impact, the sudden surge in demand for saw logs and sawn wood in Thailand, has led to widespread over-exploitation of forests in the neighbouring economies. Recently, Lao PDR announced restrictions on log exports (but not on sawn wood) to create local jobs and to increase local value addition to forest products. Myanmar has also tightened controls by raising both fees and infrastructure requirements from foreign concessionaires. The future of Cambodia's abundant forest resources is still unclear because of various uncertainties affecting sustainable management of forests.

Forest situation in thailand

In 1968, the RFD started to bring all forestlands under management plans and prepared timber-harvesting schemes. Forest management plans have been reviewed and improved several times to suit the environmental conditions and economic situations. Forest area depletion gradually decreased when the RFD intensified various forest protection measures. However, forest clearance for encroachment and illegal timber cutting still persists. Timber used to be an important export commodity and has played a significant role in Thailand's economic development. Foreign timber companies have harvested teak from natural stands in the north since the end of the nineteenth century.

Thailand used to be one of the richest forested economies in Southeast Asia. More than 50 percent of forest cover allowed Thailand to become one of the world's leading wood exporters, especially teak wood. However, the situation has changed since 1989 when logging had to be banned in response to the unprecedented flooding attributed to deforestation. This brought about a major reversal of the situation making Thailand a wood importing economy. Currently, most of the wood is being imported from neighboring economies such as Indonesia, Lao PDR Malaysia and Myanmar. During the last two decades, the situation has become worse. This was a turning point for forest management and planning resulting in the strengthening of forest protection and conservation activities. More forest lands have been declared as protected areas (e.g. national parks, wildlife sanctuaries, forest parks, and non-hunting areas). The remaining forests have been managed for protection and conservation purposes only and timber harvesting rights have been revoked.

Forest policy

Since 1961, Thailand's economic development has been guided by the National Economic and Social Development Plans (NESDP). From the first to the tenth national development plans, the concepts, approaches and strategies have evolved taking into account the changing domestic and international conditions. A significant shift in the economy's development planning took place from the Eighth Plan (1997 – 2001) onwards from a growth-oriented approach to the new model of holistic "people-centered development". During the Eleventh Plan (2012 – 2016), Thailand has continued to face major global and internal changes such as world economic crisis, population increase, global warming and environment degradation.

The National Forest Policy formulated in 1985 emphasized the need to manage the forest resources sustainably and in conformity with the development of other natural resources to enhance social, economic, and environmental benefits. The national plan aims to maintain 40 percent of the land area as forests, of which 25 percent will be for conservation and 15 percent for production.

Organizations dealing with forestry activities

The Ministry of Natural Resources and Environment (MoNRE) is responsible for the natural resources and environment of the economy. Its missions are to reserve, conserve, rehabilitate and develop the natural resources and environment with the participation and active involvement of all sectors. The strategies regarding forest resources are as follows:

- To pursue a balanced approach between conservation and utilization of the natural resources in conformity with the sustainable development approach;

- To manage the sustainable and fair utilization of biodiversity;

- To manage the water resources adopting an integrated approach; and

- To manage and develop the natural resources and improve environmental quality focusing on participation and integration at all levels.

Reforestation: past and present

The history of reforestation in Thailand began in 1906. The first teak (*Tectona grandis*) plantation of less than 1 ha was established by dibbling seeds in a shifting-cultivated area at Mae Paan Forest, Sungmen District in Phrae Province in the north, by the local forest office of the RFD. Plantations expanded to almost 160,000 ha by 1980. Up to 2000, the RFD had established 835,235 ha of forest plantations. These plantations serve two main objectives: (a) reforestation of disturbed forests, and (b) improvement of watersheds. Some areas have also been reforested mainly for amenity values (Figure 1).

Sixty years later, the Forest Industry Organization (FIO) and Thai Plywood Co. Ltd. have started their own forest plantations for commercial purposes. Teak planting has also been undertaken by the private sector during the last decade. The FIO, a government-owned enterprise, started a forest plantation program in 1967 with teak as the main species. The timber is not old enough to be harvested yet; the organization has been affected adversely by the logging ban since 1989. Hence to reduce timber imports from neighboring economies, the

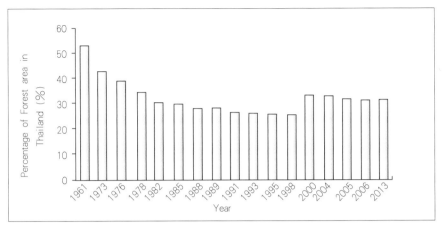

Figure 1　Change in forest area of Thailand

Source: 1961 – 2006 Forestry statistics of Thailand, 2007, Royal Forest Department; 2013 Forest Land Management Bureau, Royal Forest Department

RFD launched a forest plantation promotion project in 1994 to encourage and support private landowners and local farmers to establish forest plantations of commercial tree species covering 1.28 million ha within 12 years.

By 2000, the extent of forest plantations established under this program was still below the target. The first forest plantations of fast-growing species with rotations of 10 to 15 years (e.g. *Eucalyptus camaldulensis* and *Casuarina equisetifolia*) were established by farmers and private landowners for commercial purposes more than 40 years ago. Some commercial companies and semi-private enterprises were supported by the RFD in establishing forest plantations (e.g. the Thai Petroleum Industry, the Telecommunication Authority of Thailand and the Thai Cement Company). RFD also supports many Royal Initiative Projects to plant trees on various occasions. Nurseries managed by the RFD have been distributing millions of tree seedlings to local communities, farmers and institutions or free of cost. These trees will be an important source of timber for rural communities in the future.

By 2000, the total area of planted forests in Thailand had reached about 2.81 million ha, of which 83 percent is owned by the state and state enterprises and the remaining by private sectors. The area under different species are indicated below (FAO, 2001).

- Teak: 836,000 ha;
- *Eucalyptus* spp. : 443,000 ha;
- *Acacia mangium* and other *Acacia* spp. : 148,000 ha;
- *Pinus merkusii* and other *Pinus* spp. : 689,000 ha;
- Other broadleaved species: 541,000 ha; and

- Other conifers: 148,000 ha.

Changes in forest management

Forest resources are of utmost importance to the overall economic and social development of the economy. Protection and enrichment of forest resources for sustainable benefits, therefore, are crucial functions which involve various forestry activities. Several organizations under the overall supervision and direction of MoNRE have been assigned the responsibility for implementing various activities as stipulated in the vision and missions of the different organizations. The main functions performed by the RFD include forest protection, reforestation and rehabilitation, research and development, technology transfer, survey and research on forest biodiversity, promoting participation of local communities, community forest establishment, forest land management and the implementation of the Royal Initiative Projects. Development of collaboration with international organizations and provision of technical services also form part of the mandate of the Royal Forest Department.

Forest conservation

Forest conservation is focused on managing forest resources for sustainable benefits to people and communities and for maintaining ecosystem stability. The main activities are establishment of protected areas for forest conservation, reforestation and rehabilitation of degraded forests, research and development on forestry and related areas and promotion of people's participation in forest conservation in harmony with the lifestyle of local communities.

Forest conservation efforts in Thailand are centred on national parks, wildlife sanctuaries, no-hunting areas, forest parks, biosphere reserves, areas in watershed class 1 and watershed class 2, botanical gardens, arboretums and various experimental areas such as species, province and progeny trials (Table 1). National parks and wildlife sanctuaries are governed on the basis of specific laws and regulations relating to their protection, control and management.

Table 1 Some major forest conservation areas in Thailand

Conservation type	Number	Total area (ha)	percent of the economy area (%)
National parks	110 [1]	5,513,532	10.75
Wildlife sanctuaries	57	3,657,872	7.13
No-hunting areas	60	523,304	1.02
Forest parks	113	123,671	0.24
Botanical gardens	16	4,137	0.01
Arboretum	55	4,265	0.01

Source: Department of National Parks, Wildlife and Plant Conservation, 2008.
[1] All gazetted

Royal Forest Department has implemented a reforestation program and has established plantations of teak and other timber species annually according to the allocated budget. These plantations are established with differing objectives: commercial timber production, watershed improvement, restoration of degraded forests, environmental conservation and multiple objectives under the Royal Initiative Projects (Table 2).

Table 2   Reforestation by government organizations and plantation areas by 2007

Item	Area (ha)
Re-afforestation by government budget	723,983
The reforestation campaign in Commemoration of the Royal Golden Jubilee	455,573
Reforestation by Forest Industry Organization	43,996
Reforestation by Thai Plywood Company Limited	5,743
Reforestation according to Ministry's regulations	25,879
Reforestation by concessionaire budget	47,473
Total	1,302,647

Source: Forest statistics of Thailand, 2007, Royal Forest Department

The private sector has also been actively involved in reforestation efforts in Thailand for over 30 years. Teak, pines, casuarina, eucalyptus and acacia are the main species used for reforestation by private investors. In particular, *Eucalyptus camaldulensis* is widely preferred by private investors on account of its rapid growth.

Thailand needs to concentrate its resources to implement sustainable management policies in addition to strengthening protection programs to preserve the Kingdom's forest resources. The sustainability of plantation forestry is an issue of wide interest and concern.

Forest plantation strategy

Depletion of forest resources and logging ban have led to a decline in domestic timber supply necessitating wood import from neighboring economies. In the long run, Thailand will have to enhance wood supply from forest plantations. Ultimately, management should be geared to meet the growing timber demand through a combination of sustainable management of natural forests and enhanced efficient production from forest plantations. In the North, most important tree species include teak (*Tectona grandis*) and *Pinus* spp., in the Northeast teak (*Tectona grandis*) and *Eucalyptus* spp., and in the South and East Para rubber tree (*Hevea brasilensis*) (Table 3). Teak goes for sawn timber and eucalyptus mainly for poles and pulp industry. Para rubber trees yield latex, a critical industrial raw material and the timber is used

Table 3 Plantation areas of government organizations and private sectors in 2008

Species	Area (ha)
Tectona grandis	836,000
Pinus spp.	689,000
Casuarina spp.	148,000
Eucalyptus spp.	480,000①
Acacia spp.	148,000

① Private sectors only

for production of sawn timber and veneer.

The Royal Thai Government (RTG) has anticipated future timber shortages. Besides various forest plantation promotion strategies, the RFD has tried to encourage the private sector to invest in forest plantations. To reach this goal, research and training on the mass production of superior tree seedlings, utilizing biotechnology are essential.

Research and management

Research in forestry has been conducted mainly by the research divisions of the RFD. The main areas of research are silviculture, tree breeding and improvement, mangrove forests, forest economics, wood technology, forest insects, forest ecology, and watershed management. Effective research and development, based on appropriate genetic resources and good silviculture, are the foundations of successful plantation forestry. Many plantation forests, particularly in the tropics, are not yet achieving their full productive potential. The major research activities on reforestation have been long-term species and provenance trials, breeding for tree improvement, and silvicultural techniques. Research on tissue culture for the production and enhancement of planting materials as well as for genetic improvement has been started recently, mostly on teak (*Tectona grandis*), edible bamboo (*Dendrocalamus asper*), and *Eucalyptus camaldulensis*.

Teak plantation

Teak is one of the most valuable timbers in the world. This is due to the high quality of the wood, especially strength, durability and appearance. Teak is considered as an attractive, light, but strong wood with great resistance against fungi, humidity and insect damages. Absence of splitting, cracking and warping teak timber is found to be a user-friendly material for processing.

Teak is native to South and Southeast Asia. *Tectona grandis* tolerates a wide range of climates, but it grows best in a warm, moist, tropical climate. It prefers a dry season of 3 – 6 months which is typical in a monsoon climate. The optimal annual rainfall is 1500 – 2000 mm, but it endures rainfall as low as 510 mm/year and as high as 5080 mm/year. During the dry season, the water consumption is negligible because of its deciduous nature. Under very dry conditions teak is usually stunted. Under very moist conditions, teak is large and fluted and usually behaves like a semi-evergreen species and often in such conditions the wood quality is poor in terms of colour, texture and density (Kaosa-ard, 1981). The optimal temperature for teak is between 27 – 36°C degrees; it poorly tolerates cold and frost conditions during the winter period. According to several studies, teak requires relatively large amounts of calcium for its growth and development (Kaosa-ard, 1981). Teak may grow from sea level up to 1200 m, but growth is slower on high elevations and on steep slopes. Teak seedlings are sensitive to severe drought. Teak is also tolerant to termites, but heavy grazing by pigs, rats, deer and bison may cause damage to teak shoots. The most serious animal threat for teak plantation is the elephant. (Kadambi, 1972)

Teak plantations have been widely established throughout the tropics with the main objective of producing high-quality timber adopting rotations varying from 40 to 80 years. In general, the productivity of the teak plantation is 8 – 10 m^3/ha·year. Teak grows well on moist sites. To produce high-quality timber trees, the site should be subjected to a dry period of 3 – 5 months duration. Teak soil is deep, well-drained, and alluvial with high calcium, organic matter and other element content. The soil pH is 6.5 – 7.5. Teak is a light-demanding species. As a result, intensive weeding during the first five years is imperative. First and second thinning are conducted at ages of 5 and 10 years in close spaced plantations using a simple mechanical thinning technique.

The supply of improved seed for planting program is a major problem especially in economies where teak is an exotic. A large quantity of improved seed can be obtained through establishment and management of Seed Production Areas and Seed Orchards. Clonal propagation by tissue culture is an option for mass production of planting stock. This technique is technically and economically feasible.

There are many factors limiting the success of teak plantation establishment. Three main factors affect growth and quality of the plantation: site quality, seed supply and silvicultural management. Site is the primary factor influencing plantation growth and development.

Timber-tracking system in Thailand

Thailand is currently preparing for potential negotiations with the European Union, Association of Southeast Asian Nations (ASEAN) and other international organizations on various issues relating to the implementation of sustainable forest management. Being a leading exporter of wood products, Thailand has to take cognizance of the emerging obligations and the pressures to abide by international commitments relating to legality and sustainability. This is particularly so in the context of the government's drive to make Thailand one of the largest furniture production centers in Asia. This will require setting up standards of performance, ensuring that Thai timber products come from legal and sustainable sources. In 2010, an action plan pilot project on electronic—Timber-Tracking System was initiated jointly by the Royal Forest Department and Department of Customs in Thailand. Thailand's timber-tracking system was created using electronic equipment in order to manage the database and network information exchange for National Single Window. It has been developed based on the Chain of Custody Guidelines: Pan-ASEAN Timber Certification Initiative, Draft 2.0. 28 March 2009. The system is not developed only for facilitating imported and exported timbers and timber products but also for improving former Chain of Custody paper-based documentation system. The new system also ensures qualitative improvement of database management and strengthens Chain of Custody (CoC) information exchange with the digital system providing real-time information of the entire wood supply chain.

Timber-tracking system development is an important step in improving law enforcement and ensuring the implementation of SFM in Thai forests. The initial focus of government appears to be a technical one: to put in place the forest control technology (timber tracking) that would allow for the validation of legal timber. The challenge now is to achieve the right institutional mix to ensure a credible verification system and to establish a legal basis for the system. Considerable investment has been made to establish a national timber verification system in Thailand.

Sustainable forest management in thailand

Sustainable Forest Management is defined as a dynamic and evolving concept aimed to maintain and enhance the economic, social and environmental value of all types of forests, for the benefit of present and future generations. Currently, Thailand is highly dependent on imported timber and this will continue in the future also. It is collaborating with other economies and international organizations at bilateral, regional and international levels in implementing a number of forestry activities such as forest resources management, reforestation and forest rehabilitation, conservation of forest resources and wildlife, training and personnel development, *etc*.

Being an importer of wood and an exporter of finished products, Thailand needs to have a verifiable system of traceability that allows timber to be tracked throughout its physical movement—the wood supply chain—from the forest to the final product. To address this, Thailand has to consider the following:

- Adopting the ASEAN Regional Criteria and Indicator for Sustainable Management of Natural Tropical Forests to be as Thailand SFM;

- Establishing a special RFD unit to take responsibility for the Online Monitoring, Assessment and Reporting (MAR) Systems of Sustainable Forest Management; and

- Collecting carbon stock changes to fulfill the monitoring, assessment and reporting (MAR) requirements.

In addition, plans need to be implemented in many areas, mainly in protected area and buffer zones of community or village-managed forests. Emphasis should be on cooperative efforts involving forestry staff, local communities, villagers, and industries adopting the concepts and approaches pioneered by His Majesty the King Bhumibol Adulyadej, which have clearly indicated the scope and feasibility of rehabilitation and development of land improving the social, economic and ecological benefits.

In its efforts to strengthen sustainable forest management, the Thai government has encouraged private sector investment in forest plantation development in private lands. For example, the *Forest Plantation Act of 1992* provides incentives to owners of forest plantations by way of permitting free movement of timber produced from such plantations. Without the Act, the transportation of any logs, whether were planted or cut from private land, was once controlled strictly by the *Forest Act 1941*, acting as a disincentive in investing in commercial plantations. There are a number of private companies and individual farmers who invest in forest plantation, mostly teak and eucalyptus plantations, and many have formed Forest Farm Cooperatives to strengthen their efforts. These cooperatives receive some limited support from government, mainly in the form of planting materials and technical advice on plantation management in addition to capacity building in marketing.

Challenges ahead and the ways to move forward

Deforestation and forest degradation continue to be major problems confronting Thailand and illegal logging is a major contributory factor. Land tenure and increasing demand for land are also other factors contributing to deforestation. Policies need to be refined and improved especially to facilitate increased public participation. There is also a need for more research on participatory approaches particularly focusing on the costs and benefits, including socio-economic and environmental impacts of different options. Integrated action plans need to be continuously evaluated to ensure that actual practices on the ground are refined. There is also a need to promote better use of wood, including improving the use of wood for energy production, especially considering the role of forests in climate change mitigation. Resolution of conflicts between land tenure laws and forest laws is another area requiring urgent attention. Also promoting collaborative efforts require full involvement of local communities. Some of the lands which are owned/ occupied by poor people have been declared as protected areas and there is a need to resolve such conflicts. Sustainable forest management in Thailand is largely a human issue, especially of building trust and cooperation and if this can be accomplished, the future of forests and forestry will be brighter.

References

Department of National Park, Wildlife and Plant Conservation (DNP). 2007. Statistical Data 2005 (in Thai). Planning and Information Office, Department of National Park, Wildlife and Plant Conservation, Thailand.

DNP. 2008. Statistical Data 2006 (in Thai). Planning and Information Office, Department of National Park, Wildlife and Plant Conservation, Thailand.

FAO. 2001. Global Forest Resources Assessment 2000. In: FAO. FAO Forestry Paper 140. Rome, Italy: FAO.

FAO. 2005. State of the World's Forests 2005. Rome, Italy: FAO.

Kadambi, K. 1972. Silviculture and management of teak. Bulletin 24, School of Forestry, Stephen F. Austin State University, Nacogdoches, Texas.

Kaosa-ard, A. 1981. Teak, *Tectona grandis* Linn f., its natural distribution and related factors. National History Bulletin of the Siam Society, 29: 54-74.

Pupaiboon, U. 2012. Analysis of Timber Tracking System in Thailand (M.S. thesis). Beijing, China: Beijing Forestry University.

Royal Forest Department (RFD). 1995. Annual progress report (in Thai). Bangkok, Thailand: Private Plantation Division, Reforestation Promotion Office, Royal Forest Department.

RFD. 1996. Forest Statistics of Thailand, Data Center (in Thai). Bangkok, Thailand: Information Office, Royal Forest Department.

RFD. 2005. National Report to the Fifth Session of the United Nations Forum on Forests. Bangkok, Thailand.

RFD. 2006. Forestry Statistics of Thailand 2006 (in Thai). Bangkok,Thailand: Samlada Press.

RFD. 2008. Forestry Statistics of Thailand 2008 (in Thai). Bangkok,Thailand.

Sumantakul V, Sangkul S. 2000. The teak stump production system in Thailand: Teak Resources in Thailand.

Teak Improvement Centre. 2008. Lecture Notes and Presentation, 10.1.2008, Lampang, Thailand.

TIC. 1994. Introduction to the Teak Improvement Center (TIC). TIC, Ngao, Lampang 52110, Lampang Province, Thailand.

12 Degraded Forest Rehabilitation and Sustainable Forest Management: SFM Activities in Forest Industry Organization in Thailand

Wondee Supprasert[1]

Abstract

The Forestry Industry Organization (FIO) is a leading public sector agency responsible for establishment and management of industrial plantations in Thailand. It has a long history of managing forests for sustained yield and has embarked on more systematic efforts to manage the plantations adhering to global certification standards prescribed by the Forest Stewadship Council (FSC). Certification ensures that the plantations are managed complying the various laws, protects the environment including biodiversity and ensures that livelihood of people are taken care of. Systems that monitor impacts ensure that there is convergence of what is proposed and what is implemented and that pre-determined targets are fully accomplished.

Key problems analyzed

The urgency of rehabilitating degraded forests has become particularly important in the context of climate change adaptation and mitigation to reduce carbon emission, a significant share of which is contributed by deforestation and forest degradation. Degradation of forest ecosystems has become a major problem in Thailand.

---

1. Assistant Director of Forest Industry Organization, Thailand

Approach adopted for rehabilitation

The Forest Industry Organization (FIO) established in July 1956 is under the jurisdiction of the Ministry of Natural Resources and Environment (MNRE), Thailand. The FIO launched the programme of reforestation of degraded forests in the northern part of Thailand in 1957 adopting the taung-ya system developed in the Myanmar forests. Under the system, the local communities could plant their crops in the inter-spaces of the tree seedlings and harvest the crops. With the inputs provided for crops—fertilizers and regular weeding—the tree seedlings also grow rapidly helping in the faster rehabilitation of degraded forests.

In the context of growing environmental concerns, Thailand introduced a total ban on logging natural forests. The concept of sustainable management has been being adopted widely as a policy measure since 1987. FIO has adopted the Forest Stewardship Council (FSC) system of certification and many of the FIO plantations have been brought under the FSC certification framework. The Rainforest Alliance, founded in 1989, functions as Certification Body to certify responsible forestry practices of FSC-SFM program of FIO in 4 plantations of 19,000 ha in two provinces (Lampang and Phrae Provinces).

Adherence to FSC's regulations and principles reflects FIO's commitment to sustainable management. The certificate is effective during year 2011 – 2016. Surveillance will be carried out during the 2nd to 5th years to ensure compliance with the FSC's certification principles and requirements.

Main outcomes

The FIO-FSC forest management program has helped strengthen sustainable forest management, increasing the productivity of high-value teak plantations. Through this, the FIO aims to reduce the area that is degraded and deforested. Also the process of establishment of plantations has led to improvement in the livelihood of people in the area. SFM has also helped enhance timber supply from the plantations.

There are 16 plantations in FFF project with total area of 21,883.60 ha. Of these, 10 are under FIO and the remaining 6 are under private ownership (Table 1).

Table 1 List of FIO and private sector plantations in 2013 – 2014

FIO plantations		Private plantations	
Plantation name	Area (ha)	Name of owner	Area (in ha)
Mae Jang	3096	Mr. Kraisorn Swangdecharux	314.8
Mae Saroy	1740	Mr. Virote Theerawatwathee	800.0
Mae Hor Phra	1580	Mr. Likhit Worasiri	42.0
Mae Cham	1101	Mr. Neam Boonrom	12.8
Sri Satchanalai	2596	Ms. Porntipa Pornpanapong	80.0
Khao Krayang	2479	Mr. Wutiphong Chaisaeng	320.0
Sai Yok 1	915		
Sai Yok 2	3028		
Krerng Krawia	2659		
MaeSin-MaeSoong	1120		
Total	20314	Total	1569.6

Key element relating to the case study

In complying with the SFM standards, FIO has set forest policies under the 5 elements of the best practice of sustainable forest management in FIO plantations for the next decade. They are as follows:

- Compliance with laws;

- Safeguards of environment and biodiversity;

- Improved livelihood of local people;

- Sustainable yield; and

- Monitoring of impacts;

Reforestation will be encouraged and supported throughout the economy. This will also encompass a strong awareness programme covering households, schools, universities and other sectors. FIO will provide leadership and support for the development of business-oriented and economically viable plantations and all the 244 plantations will be certified by both FIO local standards or FSC international standards by 2020.

Summaries

The rehabilitation of degraded forest has become an urgent issue in the context of climate change adaptation and mitigation to reduce carbon emissions considering that a significant share of the emissions comes from deforestation and forest degradation. Complying with the SFM standards, FIO has set forest policies under the 5 elements of the best practice of sustainable forest management in FIO plantations for the next decade. SFM has become the basic principle for management for all the FIO plantations helping improve the environment, especially through carbon sequestration and enhancing the livelihood of local communities.

13 An Overview of Forest Rehabilitation in Vietnam

Nguyen Tuong Van[1]

Abstract

The paper outlined the efforts being pursued by Vietnam to rehabilitate its forests and discussed the underlying causes of deforestation and degradation and what had helped reverse the situation. Vietnam has been able to arrest deforestation and move on to a path of forest transition through concerted efforts including reform of policies and legislation, strengthening institutional framework and improving the science and technology base. Strong commitments at the political level have created a very favourable environment, enabling the involvement of all key stakeholders. Vietnam has articulated a well thought-out National Forest Development Strategy with several coordinated activities. Coordinated activities, which has a time horizon of 2020 The paper provided an analysis of the performance of the forest development strategy as also the factors that have contributed to the success.

Introduction

Vietnam has a land area of over 33.12 million ha, of which 12.6 million ha of forests and 6.16 million ha of barren land are targeted for agriculture and forestry production. The forestry sector has been managing and running production activities on the largest area of land as compared with other sectors in the national economy. The forest area is distributed mainly in the mountainous and hilly areas, where 25 million people from different ethnic groups live.

---

1. Forest Ranger of Propaganda and Personnel Development Division of Forest Protection Department, Vietnam

These people are highly dependent on natural resources, in particular forests, and face a number of livelihood problems considering the less developed condition of the area where they live.

The state of forests

Due to unsustainable management and the very high demand for land for cultivation and for forest products, the forest area and forest quality have been continuously declining during the last few decades. In 1943, Vietnam had 14.3 million ha of forests or about 43 percent of the land area; by 1990 this has declined to 9.18 million ha or 27.4 percent of the land area. Average annual loss of forests during the period 1980 – 1990 was estimated as more than 100,000 ha. But from 1990 onwards, the situation has reversed and the forest area has increased gradually due to afforestation and rehabilitation efforts (Figure 1). In 2006, the total national forest area was 12.61 million ha (forest coverage of 38 percent), including 10.28 million ha of natural forests and 2.33 million ha of forest plantations, which can be classified into the following three types.

- Special-use forest: 1.93 million ha;
- Protection forest: 6.20 million ha;
- Production forest: 4.48 million ha.

Government programs have played a key role in increasing the forest cover. Yet the forests in Vietnam are under serious threat especially in some regions like Central Highlands, the Central Coast and the Southeast Region. Further, forest degradation and fragmentation

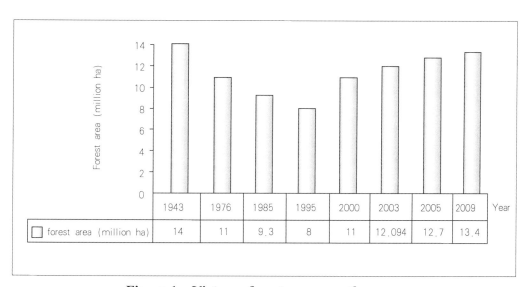

Figure 1   Vietnam forest area over the years

remain major problems throughout the remaining natural forests. Over two-thirds of Vietnam's natural forests are considered to be poor, while the rich and closed canopy forest constitutes only 4.6 percent of the total forest area. Between 1999 and 2005, the area of natural forest classified as rich decreased by 10.2 percent.

With these forest resources, the per capita forest area is only 0.15 ha and the per capita growing stock of timber is 9.16 m^3. This is far below the global per capita forest area of 0.97 ha and the per capita growing stock of 75 m^3, making Vietnam a forest poor economy.

Vietnam has about 6.76 million ha of barren unused land, most of which is in the hilly and mountainous area. Seventy one percent of this is in areas where elevation is less than 700 m. Potentials of these lands remain unrealized on account of a number of unfavourable factors like steep slopes, degraded status and scattered distribution.

Causes of deforestation in Vietnam

The causes of forest cover decline are complicated and diverse, and widely debated. Some of the important factors contributing to deforestation are as follows:

- Land conversion for expansion of agriculture;

- Devastation by war, including two anti-invasion wars, from 1945 to 1954 and 1961 to 1975. During these wars, Vietnam lost nearly 2 million ha of forests;

- Forest fires;

- Overharvesting of fuelwood and timber by state organizations as also widespread illegal logging; and

- Poor management capacity of the forestry sector and a deficient institutional and legal framework.

Forest rehabilitation efforts

Vietnam is known for its efforts to rehabilitate its forest cover, in addition to its drive to develop its forestry, and wood and non-wood forest product-based industries. Initiatives for rehabilitation of forests are summarized below.

Scattered tree planting

A significant contribution to Vietnam's forest rehabilitation is the scattered tree planting initiative. This initiative began in the 1950s and was endorsed by Chairman Ho Chi Minh in 1959 as the tet tree planting festival. It has led to massive tree-planting by people as an annual event and several billion trees have been planted since 1955. An estimated 1 to 2 billion trees, equivalent to 1 to 2 million ha have been planted in the economy.

World Food Programme

Another significant forest rehabilitation initiative was supported by the World Food Programme (WFP). WFP implemented 6 forestry projects, with a total budget of US $ 160 million. These funds were used to supply food to rural communities, provide equipment and materials for 450,000 ha of forest plantations, construct forest roads, organize fire protection measures and improve forest extension services. The WFP forestry projects were focused on the development of demonstration plots and agroforestry production on steep slopes. Farmers were given the opportunity to select their own crops and species to plant.

The WFP forestry projects have yielded good results. Extensive areas have been planted creating employment for a large number of people and improving livelihood of rural communities. Also this has helped develop new forest plantation and agroforestry techniques, and enhance gender equity in forestry and local staff have been trained in the organization and management of forestry projects.

Rehabilitated forests at the beginning of the large programs

Since the early 1990s, following the United Nations Conference on Environment and Development (UNCED) and initiatives like the Tropical Forestry Action Plan (TFAP), Vietnam has embarked on major reforms in the sphere of natural resource management. The major forest rehabilitation programs implemented are:

- Greening the barren hill program (Program 327); and
- The five million hectare restoration project (Program 661).

The first, greening the barren hill program, was completed during the 1990s while the five million hectare restoration project commenced in 1998 and was completed in 2010.

Vietnam forest development strategy and forest rehabilitation projects

Vietnam Forest Development Strategy 2006 – 2020

Currently, forestry sector activities in the economy are based on the National Forest Development Strategy (NFDS) 2006 – 2020. It builds on the previous strategies and programs, setting out ambitious targets for policy reform, plantations, financial support for forest protection and plantations and a greater role and responsibility for the local communities. It seeks to modernize the forestry sector, so that the sector can play its part in the industrialization and modernization of rural agriculture, in hunger eradication, in poverty alleviation of people in the mountainous areas and in protection of the environment. The NFDS gives particular thrust to the need for clear ownership of land and forests. It also discusses the enforcement of land laws, providing guidance on the responsibilities of local communities and other stakeholders.

Objectives

- To sustainably establish, manage, protect, develop and use 16.24 million ha of land set aside for forestry;

- To increase the forest area to 42% – 43% of the land area by the year 2010 and 47% by 2020 ensuring wider participation from various economic sectors and social organizations;

- To increase forest's contribution to socio-economic development, environmental protection, biodiversity conservation and environmental services supply;

- To reduce poverty and improve the livelihoods of rural mountainous people; and

- To contribute to national defense and security.

Solutions

The above policy objectives are proposed to be accomplished through the following measures.

- Solutions on policy and laws.
  - Policies for forest and forest land management; and
  - Finance and credit policies.
- Renovation of organizing forest production and business and encouragement of economic entities participating in forest protection and development.
- Improved planning, plan and monitoring.
- Sector organization and management.
- Science and technology.
- Human resources development.
- International cooperation.

Programs

Vietnam is implementing three development programmes and two support programmes to accomplish the development strategy as indicated below.

Development programmes

- Sustainable forest management and development programme;
- Forest protection, biodiversity conservation and environmental services development programme; and
- Forest products processing and trade programme.

Support programmes

- Research, education, training, and forestry extension programme; and
- Renovation of the forestry sector institutions, policies, planning and monitoring programme.

Forest rehabilitation projects in vietnam

Forest rehabilitation is undertaken in all the three broad categories of forests (Protection Forests, Production Forests and Special-use Forests) in order to fulfill the objectives expected from each of the specific forest type. The types of projects implemented are:

- Protection forest rehabilitation projects;
- Special-use forest rehabilitation projects; and
- Projects on production forest land (which includes production of raw material for pulp and paper including wood chips, and valuable wood from indigenous species).

In addition, there are projects supporting forest rehabilitation, including those aimed to provide technical assistance, seed production and social forestry projects.

Objectives of rehabilitation projects

In Vietnam, each of the three forest types have a specific objectives: protection forests to protect upstream areas of watersheds or control sand movements in coastal formations; special-use forests to conserve natural or cultural heritage and production forests to supply forest products. However, considering the nature of demand on the forests, often projects aim to accomplish multiple objectives and most rehabilitation efforts aim to fulfill production, protection and other functions taking into account the specific characteristics of forests in each area. Broadly, the objectives of rehabilitation/restoration projects can be grouped into the following categories:

- Catchments protection/biodiversity conservation;

- Restoration of forest cover/re-greening;
- Poverty alleviation, rural development and employment generation;
- Promotion of tourism;
- Production of wood and other products;
- Knowledge and technology creation; and
- Others.

Forest rehabilitation projects are executed by the following agencies:

- The Department of Agriculture and Rural Development;
- Forestry agency;
- People's Committees;
- Management boards/project management units;
- State forest enterprises;
- Donors;
- Science agencies/ institutions; and
- Others.

Funding

Forest rehabilitation projects are funded either nationally (which includes state budget, loan, direct investment or self-financed) or internationally (technical assistance funds, loan or joint venture). Some of the specific issues relating to funding are outlined later.

Method

A wide array of approaches are adopted to rehabilitate degraded forests, including natural regeneration, natural regeneration plus enrichment, enrichment, protection, plantation with natural regeneration, planting and agroforestry including intercropping.

Challenges

Forest rehabilitation in Vietnam faces a number of challenges as indicated below.

- Population growth is resulting in spontaneous migration and the inefficient land use in mountainous area is putting enormous pressure on forests for expansion of agricultural land.

- Increasing demands for forest products have put pressures on forest resources and the environment, particularly on natural forests. The current demand for forest products far exceed the sustainable supply from the forests. Land that could produce high yields of wood is limited and occurs in scattered patches limiting their potential for industrial wood production.

- The competitiveness of forestry production is still low. International integration not only is an opportunity, but also a great challenge for the forest product processing industry and trade of forest products. The competition will be more critical in the future, in both international and domestic markets.

- Resources available in the forest sector (e.g. human resources, infrastructure, funding, management capacity, *etc*.) are inadequate to meet the needs for comprehensive and sustainable development of the sector.

- The importance of forestry has not been comprehensively, objectively and fairly evaluated, which has affected the formulation of investment and development policies in the sector.

Outcomes of forest rehabilitation projects (up to 2005)

Forest cover—conservation achievements

- Forest cover reached the proposed objective.
- 43,000 ha replanted.
- Planting forest, and improving landscape quality (1,350 ha).
- Restoration of barren lands in special-use forests.
- 3000 ha of acacia planted between 1992 and 2003.
- Reforestation, tending and protection.
- Core areas of special-use forests are well protected.
- Selected appropriate species for dry and coastal areas.
- Cutting of natural forests reduced.
- Annual monitoring of biodiversity.

Social achievements

- People's participation in reforestation improved.
- Resettlement and training.

- Training to transfer reforestation technology in alkaline soil for local staff and farmers.
- Farmers receive stable prices for wood.
- Improvements in education, health and culture.
- Lives of local people improved.
- Local people participating to formulate plan and carry out forest rehabilitation and utilize forest.
- Assistance to ethnic groups to leave protected areas.
- Productivity achievements.
- Stable supply of raw material for Vietnam Paper Corporation and to VIJACHIP (Vietnam Japan Chip Corporation Ltd—an afforestation and woodchip production business company).
- Enhanced productivity and improved quality of forest.
- Technology for rehabilitation improved.
- Selection of species that are appropriate and of high economic value in alkaline soil in the Cuu Long Delta Region.
- Recommended solutions to improve alkaline soil and protect water sources in the course of reforestation process.
- Promotion of ecotourism.

Summaries and lessons learnt

The results of 50 years of forest rehabilitation

Vietnam has a long history of forest rehabilitation, as it started to give tree planting its due importance in the mid-1950s. Since then, the economy has made a great deal of effort on bringing back tree vegetation where forests have disappeared.

There are various indicators that demonstrate the success of Vietnam's forest rehabilitation. Over 85 percent of the projects that were surveyed as part of this report have met their main and specific objectives. The people who provided success ratings of their projects rated over 50 percent as successful or good, while over 80 percent of the projects rated between quite successful and very successful.

Most of the forest rehabilitation projects included in the surveys had more than one objective. The objectives related to restoration of forest cover for productivity, environmental functions including biodiversity conservation, but also to local and wider development objectives. The multiple objectives are a common feature in the majority of projects despite the dominance of forest rehabilitation on protection forest land.

Project achievements fairly well matched the objectives. Besides, there is an additional outcome of some of Vietnam's forest rehabilitation projects, as they have improved dialogue between the authorities and other stakeholders.

A different indicator that reflects the success of Vietnam's forest rehabilitation is the relation between areas rehabilitated and the existing area of forests at various points in history. Plantation forest area has increased markedly year by year: 1.050 million ha in 1995; 1.471 million ha in 2000; 2.218 million ha in 2004; 2.219 million ha in 2009; 3.438 million ha in 2012. There is no indication of how much forest has been rehabilitated naturally until the early 1990s as a result of active protection or abandonment. However, it is obvious that maintenance and assisted natural regeneration of depleted forests, especially in watershed areas, has contributed to the increase in forest coverage. Up to the year 2005, assisted natural regeneration had been applied in an area of 723,450 ha of forest under the 5 Million Hectare Reforestation Plantation (5MHRP) (accomplishing 72 percent of the plan). In 1995, forest covered only 28 percent of the economy, but this figure increased to 35 percent, 36.7 percent, 39.1 percent, 40.7 percent in 2000, 2004, 2009 and 2012, respectively.

According to the Ministry of Agriculture and Rural Development report on the achievements of the 5MHRP, until 2010, an area of 2.45 million ha of plantation forests had been established, of which 631,317 ha was protection and special-use forests, 683,396 ha was production forest, and 86,954 ha was fruit tree and industrial crop plantation on forestry land.

Forest plantations have progressively contributed to wood supply in Vietnam. In 2001, the forestry industry consumed 1.6 million m^3 of forest plantation wood. It is difficult to establish how much forest rehabilitation has contributed to enhancing forestry's share in the national GDP. With an area of over 2 million ha, special-use forests, including national parks and natural reserves, have huge advantages in biodiversity and gene conservation.

There are several positive outcomes of forest rehabilitation for local communities as they form an important group of beneficiaries by way of benefiting from cash income, non-cash incomes, forest protection contracts, use of fuelwood and non-timber forest products and opportunities for educational improvements. There are also positive environmental outcomes on account of forest rehabilitation projects in terms of improved floristic diversity, landscape diversity, soil quality and reduction of soil erosion. Many projects have contributed to forest recovery.

Explaining the outcomes

Policy and legislation

The policy and legislation in Vietnam has been highly conducive to forest rehabilitation. The Government of Vietnam has made forest rehabilitation a priority since the mid-1950s, and this commitment has been strengthened since the early 1990s. The policy of forest rehabilitation has been clearly reflected in the various national projects with consistent long-term support over many years. The rehabilitation projects have particularly taken into account the protection function of forests, especially conservation of biodiversity and protection of water and soil.

The successful results of forest rehabilitation also depend greatly on sectoral and non-sectoral policies. Invariably, all policies relating to land ownership, incentives, land use planning and environmental services all affect rehabilitation efforts. Particularly relevant have been the revisions of the *Land Law* (in 1993, 1998, 2000, 2003 and 2013) and the enactment of the *Forest Protection and Development Law*. The *Land Law* clearly states that the land is planned and generally managed by the state, but can be allocated to individuals, households, social organizations and communities for long-term use in compliance with agreed purposes. Rights are quite comprehensive as owners can exchange, transfer or inherit land use rights, or use the land as collateral for bank loans. The *Forest Protection and Development Law* provides the legal framework for forest land allocation and the leasing of forests to individuals, households, management boards, economic organizations and communities. The law provides the framework for investing in, encouraging and supporting forest protection and development, expanding the market for forestry products and insuring forest plantations. There are also a number of decrees and decisions issued by the government regarding land allocation and forest contracting, credit policies for forest protection and development. Many policies have been endorsed and amended to make them consistent with the actual situation.

National policies and legislation are being adjusted to reflect new opportunities and needs. The *Environment Protection Law*, for instance, was revised in 2005, as was the *Forest Protection and Development Law*, to take better cognizance of the role of forests in the provision of environmental services and to open opportunities for compensation where these services are being provided.

Funding

Vietnam has for many years invested a significant amount in forest rehabilitation, especially since the 1990s. This national investment has been complemented with substantial international support.

Under the current arrangements of payments for the protection of forests, state financing of forest protection needs to continue to ensure the effective accomplishment of the objectives. There is little other funding for forest rehabilitation, especially for the rehabilitation of production forest land that is meant to boost the forestry sector's contribution to the national economy.

This funding situation does not translate to optimal conditions for smallholders. Some payments, such as for forest protection contracts, are perceived to be too low. Credit available for forest rehabilitation needs to be more favourable; but even when credit availability is not a problem, convincing farmers to invest in forest rehabilitation remains challenging.

Objectives of rehabilitation

Vietnam's forest rehabilitation efforts aim to accomplish environmental, economic and social objectives and most often the projects are able to accomplish multiple objectives. Productive objectives can be carried out on production forest land, and in principle these objectives are compatible with social objectives, like improving the well-being of the rural poor. There are however many challenges in balancing the different objectives. Prices paid for wood and timber are limited by profit margins, and they may be too low to be attractive to small-scale tree growers. Local markets for wood or other forest products may be limited. Large commercial enterprises may not have much interest in dealing with many small producers. These constraints diminish the compatibility of various objectives of forest rehabilitation.

The objectives of forest rehabilitation are relatively flexible and can be adjusted if needed. For example, as mentioned above, considerations are currently being made regarding narrowing the area of protection and special-use forest, and expanding the area of production forest.

Economics, markets and demand

Various issues relating to economics, markets and demand have already been outlined in the earlier paragraphs. The woodchip and derivatives sector may suffer from high production costs, in which case nationally produced products may become less competitive in comparison with what is produced elsewhere. New product development will be an important aspect that needs to be addressed if the planned expansion is to be successful, and to make rehabilitation on production forest land economically viable. Forest rehabilitation focused on environmental functions is unlikely to become profitable any time soon, while some of the anticipated benefits from forest rehabilitation may not be realized because of the unclear link between forest cover and downstream flooding, or limited water volumes.

Recently, although forest plantation has helped increase the supply of industrial wood for production of paper, fibre and particle board, and woodchips, the demand remains large. Demand supply gap is particularly wide in the case of furniture industry whose rapid growth has resulted in increased dependence on wood imports which have currently met 80 percent of the wood demand. This implies the continued need for expanding the area under plantations for industrial wood production.

Technology, extension, technical assistance and training

Various policy makers on the Vietnam's forestry sector have observed several technical limitations to forest rehabilitation, including inadequate seed material, poor soils in plantation sites, and inadequate plantation maintenance. It should be acknowledged that science and technology, as well as the application of advanced techniques in production, have contributed significantly to improve forest rehabilitation efforts in Vietnam. Already considerable research has gone into identifying high productivity species that are economically and environmentally valuable, and able to grow on degraded/difficult sites. Advanced methods in terms of tree improvement, intensive afforestation, productivity increases, and site selection have been widely applied in the field. Good results for natural forest rehabilitation through maintenance, assisted regeneration and enrichment planting have also been achieved through the application of techniques based on relevant research. However, the need to improve productivity and to improve the supply of high quality tree seeds/planting materials remains important.

In this regard, substantial attention has been paid to improve forestry extension services. Agriculture and forestry extension organizations, as well as other governmental extension programs have been established at all levels from local to central. A number of projects for agriculture and forestry extension have been implemented. However, the effectiveness of extension service requires considerable improvement.

Lessons learnt

The following key lessons can be synthesized from the success and shortcomings of Vietnam's forest rehabilitation.

- Forest rehabilitation should be incorporated in projects and programs at the national level and implemented through projects at the local level with well-defined goals. The more detailed the project objectives and plans of operations are, the better will be the accomplishments on the ground.

- Streamlining the procedure for project appraisal, management and monitoring of project operation is essential to ensure the success of projects.

- Clear and detailed benefits for households and articulated participation will vastly enhance project results.

- Clarifying land ownership conditions for the key stakeholder responsible for implementing rehabilitation, and adequately addressing technical requirements, will also enhance project results.

- The implementation of forest rehabilitation projects should be integrated with other projects that aim to improve the socio-economic conditions of local populations.
- Forest rehabilitation projects should be combined with other supporting activities to ensure that the major goals of the projects are met.

References

Dinh Sam Do, Hung, Van Hung Trieu, et al.. 2004. How does Vietnam Rehabilitate its Forests.

Forest Inventory and Planning Institute. 1995. Investigation, Evaluation and Monitor of Forest Resource Changing in the Whole Economy 1991 – 1995. Hanoi, FIPI.

Forest Protection Department. 2014. Statistical Data of Forest Succession.

Government of Vietnam. 2005. National Report to the Fifth Session of the United Nations Forum on Forests.

International Center for Environmental Management. 2003. Vietnam National Report on Protected Areas and Development. In: International Center for Environmental Management. Review of Protected Areas and Development in the Lower Mekong River Region.

Lamb, D. and Gilmour, D. 2003. Rehabilitation and restoration of degraded forests.

Ministry of Agriculture and Rural Development. 1998. Evaluation Report of Technical Issues of 6-year Implementation of 327 Program. Hanoi: Ministry of Agriculture and Rural Development.

Ministry of Agriculture and Rural Development. 2001. Five Million Hectare Reforestation Program and Partnership: Synthesis Report. Hanoi: 5MHRP Partnership Secretariat, International Cooperation Department, Ministry of Agriculture and Rural Development.

Ministry of Agriculture and Rural Development. 2004. Forest Land and Unused Land for Forestry Planning. Hanoi: Ministry of Agriculture and Rural Development.

Ministry of Agriculture and Rural Development. 2005. Conclusion of Deputy Prime Minister Nguyen Tan Dung in the meeting of evaluation of the implementation of 5 million hectare forestation project and forest protection. Science and Technology Journal of Agriculture and rural Development.

UN-REDD Phase II Programme: 2013. Operatinalising REDD + in Vietnam.

Vietnam Forestry Development Strategy 2006 – 2020 (promulgated with the Decision No. 18/2007/QD-TTg, dated 5 February 2007, by the Prime Minister).

14 Restoration and Sustainable Management of Forest Ecosystem in the Central Highlands of Vietnam

Nguyen Tuong Van[1]

Abstract

This paper provided an overview of the ongoing efforts to restore and rehabilitate the forest ecosystem in the Vietnam Central Highlands. The forests in the Central Highlands are one of the richest biodiversity area in Vietnam, hosting plant families from diverse bio-geographical regions in Asia. However, it is also one area subjected to intense human interventions causing deforestation and degradation. The project "Restoration and Sustainable Management of the Forest Ecosystem in the Central Highlands in the Period 2013 – 2020, Vision 2030" aims to undertake a series of integrated activities enabling the systematic rehabilitation of degraded areas and promoting sustainable forest management and ensuring the involvement all key stakeholders. Improving the livelihood of local communities is a basic objective of the project. The paper also listed a number of recommendations to improve the effectiveness of the project.

Introduction

The Western Highlands in Vietnam (sometimes also called Central Highlands or Midland Highlands) consists of the provinces of Dak Lak, Dak Nong, Gia Lai, Kon Tum and Lam Dong with a geographical area of 5.46 million ha. Its unique geography, topography, climate and soil have created forest ecosystems with high floral and faunal diversity. Western Highlands is the home of multiple plant families from diverse biogeographic regions in the World:

---

1. Propaganda and Personnel Development Division of Forest Protection Department

- Dipterocarpaceae from Malaysia – Indonesia;

- Combretaceae from Indo-Myanmar region;

- Lythraceae from the Himalayas;

- Pinaceae from Yunnan – Guizhou; and

- Fagaceae, Orchidaceae and Asteraceae from Northern Vietnam.

Western Highlands has over 224 families 1200 genera 4.500 plant species. There are over 700 tree species belonging to 90 families, representing both gymnosperms and angiosperms.

However, forests in the region are in a highly degraded state due to many different causes. The current main direct causes of degradation are the conversion of forestland to agriculture (particularly to industrial crops), unsustainable logging (notably illegal logging), agricultural expansion by people migrating to forested areas, infrastructure development and forest fires. Natural forests, especially the dry dipterocarp forests have been cleared for raising industrial crops and the pace of decline of dry dipterocarp and broadleaved evergreen forest ecosystems is alarming.

It is to address the above challenges that the project "Restoration and Sustainable Management of the Forest Ecosystem in the Central Highlands in the Period 2013 – 2020, Vision 2030" was initiated. The project aims to restore the degraded forest ecosystem, thus contributing to the sustainable socio-economic development of the Central Highlands enhancing national security.

Current status, objectives and tasks

The current status of protection and management of the forest ecosystem

As per official statistics (Decision No.1739/QD-BNN-TCLN dated 07.31.2013 of the Minister of Agriculture and Rural Development), the extent of forests in the Western Highlands is 2.9 million ha or about 53.11 percent of the land area. During the five-year period of 2007 to 2011, the Western Highlands had lost about 129,686 ha of forest land (consisting of 107,425 ha of natural forests and 22,261 ha of forest plantations) due to conversion of forestland to other uses, illegal encroachment and logging. Along with the loss of the forest area, the forest quality (especially the quality of natural forests) had also declined significantly. The extent of good-quality forests with high growing stock is only 16 percent of the natural forests, mostly consisting of special-use forests. Forests recently restored through natural regeneration are relatively young with limited ability to meet production and protection functions.

The Western Highlands form a hotspot of violation of forest protection laws in the economy and during 2008 – 2012, there have been over 8600 cases of illegal deforestation reported from the region. Evidently, forest management suffered from several weaknesses. Most forest companies are operating inefficiently and are on the brink of bankruptcy. Establishment of a large number of wood processing units without any consideration of stable wood supply has led to illegal logging. Institutional inadequacies have undermined the effective protection and management of forests and even nature reserves and national parks are subjected to logging.

Restoration and rehabilitation of western highlands forests

Objectives of restoration

The main objectives of restoration and rehabilitation of forests in Central Highlands are:

- To protect and restore the existing forest ecosystem and manage the land and forest resources effectively and sustainably;

- To increase the forest cover to 55 percent of the land area by 2020 and increase the productivity, quality and overall value of forests; and

- To create jobs, promote forest-based livelihoods, contribute to hunger eradication and poverty reduction and ensure security and defense.

Vision 2030

Restoration and rehabilitation of the Western Highlands are guided by the following vision adopting a time frame of 2030:

- To protect, restore and sustainably develop 3.113 million ha of the existing forests;

- To continue to protect unoccupied land area with regenerating trees to increase the forest area; protect genetic resources of rare and valuable plants and animal species and forest ecosystems; encourage all individuals and organizations and economic sectors to participate in forest protection and management; strengthen forest management including through forest fire prevention ensuring that commune administration is fully involved;

- To maintain around 25 – 30 percent of the existing natural forest area as protection and special-use forest land, and about 10 percent of the existing natural forest area as production forest land;

- To improve the productivity and quality of natural production forests and sustainably manage them to produce an average growth at 5 – 6 m^3/ha/year by 2020;

- To continue to manage the protection and production forests to increase the forest cover, and continue to afforest the watersheds of large river basins to reduce the incidence of landslides and flash floods;

- To plant special-use forests, primarily plant native trees to renovate landscapes and to upgrade the botanical gardens;

- To afforest the forestland converted to other uses, such as farms in hilly land;

- To continue to restock forests after logging; restore poor forests and young forests and facilitate the growth and development of regenerating trees; and

- To promote tree planting by organizations, agencies, schools, local communities and households.

Tasks

The following are the key tasks being implemented during the period 2013 – 2020 enabling the accomplishment of the Vision 2030.

Protection of forests

The proposed tasks include:

- To protect, restore and develop sustainably 2.889 million ha of forests (Dak Lak Province—640,900 ha; Dak Nong Province—285,200 ha; Gia Lai Province—719,800 ha; Kon Tum Province—665,200 ha; Lam Dong Province—578,300 ha);

- To reduce the violations of forest protection and management laws, improve the efficacy of implementation of protection functions to better contribute to biodiversity conservation and thus accelerate the pace of sustainable socio-economic development of the economy;

- To protect and regenerate unoccupied lands to increase the extent of forests; Strictly protect regions of genetic resources of rare and valuable plant and animal species; and

- To encourage individuals, organizations and all economic sectors to participate in the protection and management contracts.

Development of forests

- Increase the productivity and quality of natural production forests with 1.6 million ha of permanent forest stand. Sustainably manage the forest stand to stabilize productivity at 4 to 5 m^3/ha·year by 2020.

- Increase the productivity of existing plantations and natural forests. Plant 47,815 ha of production forests to increase the production forest area to 300,890 ha, including 220,000 ha suitable for logging with an average felling cycle of 12 years and an average yield of 10 m^3/ha. Establish raw material zones providing stable and sustainable wood supply to wood processing enterprises.

- Plant 9,342 ha of protection forests in the watersheds of large rivers, lakes, hydropower and irrigation dams, and border corridor regions, especially on critical flood prone areas.

- Nurture forests after logging; nurture poor forests and young forests restored after hill farming; facilitate the growth and development of generating trees.

- Regenerate annually 14,669 ha of special-use and protection forests on unoccupied lands.

- Improve natural production forests to plant 100,000 ha of economic plantation on the poor dipterocarp forestland.

- Encourage and support the movement of scattered tree planting in organizations, agencies, schools, local communities and households.

Contents and solutions

Contents of "Restoration and Sustainable Management of the Forest Ecosystem in the Central Highlands in the period 2013 – 2020"

Forest protection

- Forest protection contracts: in the period 2013 – 2020, contract 20 percent of the natural forest area for improved protection by communities and other stakeholders.

Forest development

- Improve the productivity and quality of natural production forests with 1.6 million ha of a permanent forest stand. The objective is to sustainably manage the forests to reach an annual growth rate of 4 – 5 m^3/ ha by 2020.

- Increase the productivity and quality of 302,091 ha of existing plantations (which includes 253,075 ha of production forests); plant an additional 47,815 ha of production forests so that the total extent of production forest plantations to 300,890 ha.

- Plant protection forests to cover an area of 29,677 ha in three phases.

- Plant special-use forests: give priority to plant native tree species to renovate landscapes and upgrade botanical gardens.

- Nurture forests after logging; nurture poor forests and young forests restored after hill farming; facilitate the growth and development of generating trees.

- Regenerate naturally 14,669 ha of special-use and protection forest on unoccupied lands during the period 2013 – 2020, including 5,421 ha of special-use forest and 9,248 ha of protection forest per year.

- Improve natural production forests for economic plantation. In the period 2013 – 2020, improve natural production forests by planting 100,000 ha of economic plantation on the poor dipterocarp forest land in the whole Western Highlands region.

- Enhance the movement of scattered tree planting by organizations, agencies, schools, local communities and households.

Synthesis of investment and labor demand

The total investment required during the period 2013 – 2020 is estimated at about 6,153.3 billion Vietnam dong (or approximately US $ 300 million). Of this, 789 billion dong, accounting for 12.8 percent of total investment, will be provided by the state budget. The average capital investment of the whole region will be about 879 billion dong/year.

To perform the project efficiently, the demand for labor will be approximately 800,000 employees, or about 23,000 employees per year per province.

Assessment of the effectiveness of the project

Impacts of the project

The implementation of the project "Restoration and Sustainable Management of the Forest Ecosystem in the Western Highlands in the period 2013 – 2020, Vision 2030" will have significant short-term and long-term impacts. It will positively change the whole face of the forestry sector in the Western Highlands. Besides, it will positively impact the livelihoods of local people. At the same time, the project will also strengthen national defense and security system, maintain social order, improve food security and fulfil several socio-economic and environmental objectives. The project provides a unique opportunity to enhance the role of forestry in the socio-economic development of the region.

Economic efficiency

Implementing the restoration and sustainable management project will increase the direct benefits from forests, and the forest cover is expected to increase to 55 percent of the geographical area by 2020. In addition to improving production of wood, it will also increase the scope for increasing income through payment for ecosystem services, thereby contributing to hunger eradication and poverty reduction.

The production forest area will increase to 300,890 ha, of which 220,000 ha of forests could be logged on a cycle of 12 years yielding a potential output of 10 m^3/ha·year. GDP per capita will reach at 20.7 million dong by 2015, equivalent to US $ 1,080 – 1,110, and approximately 38.5 million dong by 2020, equivalent to US $ 1,930 – 1,950.

The effect on the environment

The effective development of natural forests, regeneration forests, forest plantations and scattered tree systems in urban areas and industrial zones, and the establishment of agroforestry system will improve watershed protection, prevent soil erosion and help mitigate floods and other such disasters caused due to land degradation. Moreover, it will renovate landscapes of urban areas and industrial zones and enhance the potential for ecotourism. Besides, it will enhance national security and preserve ecological environment.

The effect on society

- The project will contribute to the socio-economic development of remote areas and areas with special difficulties, and mobilize labour for agroforestry sectors. This annual labour mobilization will create jobs for the ethnic minorities living close to forests.

- Estimated employment in the fields of afforestation, forest rehabilitation, forest protection and management will account for 85 percent of the total number of workers; harvesting and processing will account for the remaining 15 percent of the employment.

- The average income of local people from agroforestry production activities will increase by 30 – 35 million dong/ha·year.

- Involvement of a several stakeholders in the management of natural forest resources (state-owned enterprises, private enterprises, organizations, communities, households, *etc.*) will strengthen the socialization of the forestry sector and curtail the loss of forest resources.

- The investment in infrastructure and services such as the construction of roads, agroforestry and other farming systems, and production of industrial wood will create favourable conditions for local people to improve their forestry skills and enhance access to markets. For this reason, it will help bring about positive change in the socio-economic conditions in the mountainous rural areas.

- The creation of stable incomes and creation of awareness about forest policies will consolidate the knowledge and experience of local people, especially the ethnic minorities who live close to forests, enabling their active participation in the forestry activities, and contribute to social order and safety and political security in the provincial areas.

- Awareness of the rehabilitation and sustainable management of forest ecosystem of privates, organizations and households will be strengthened towards the regulation "the more efforts of forest rehabilitation and protection, the more benefits one earns".

Implementation arrangements

The responsibility for implementing the project is shared by several institutions as indicated below.

People's Committee of provinces of the Western Highlands

- After the approval of the project "Restoration and Sustainable Management of the Forest Ecosystem in the Western Highlands in the period 2013 – 2020, Vision 2030", Provincial People's Committees will be responsible for inviting domestic and foreign investors to participate in the project.

- Guide the implementation of objectives, tasks and contents of the project under its schedule and current regulations of the state.

- Conduct reviews and adjustments of the project of restoration and sustainable management of the forest ecosystem corresponding with the zoning orientation.

- Continue to undertake research and propose mechanisms and regional, local, sector policies to authorities for approval to accomplish project objectives; monitor, verify and supervise the implementation of the regional projects in the sectors and the local.

Departments of Agriculture and Rural Development of provinces

- Develop, synthesize and propose project implementation plans to People's Committee of provinces.

- Organize, direct and supervise the implementation of development plans of the projects of provinces.

- Provide support to Boards of Project Management.

- Organize the implementation of approved programs and plans of the project.

Recommendations

- Undertake and complete forest inventory to facilitate the establishment, review, adjustment and clear determination of stable stands throughout the Western Highlands.

- Organize the implementation of the project of restoration and sustainable protection of forest ecosystems in the Western Highlands.

- Concentrate on protection, regeneration, plantation and enrichment of forest ecosystems, especially poor and degraded plantations; convert the plantations without abilities of restoration to other uses to maintain the function of environmental protection and improve livelihoods.
- Evaluate the forest planning process, particularly as regards the dipterocarp forest ecosystems with specific attention on rehabilitation and reforestation of rubber tree plantations on the forest land.
- Improve collaboration between the provinces of Western Highlands and Vietnam Rubber Group for the development of sustainable rubber plantations.
- Implement the project for improving exploitation of natural forests and for unifying the implementation of forest plans.
- Continue the renovation and consolidation of forest companies at various levels—ministries, sectors and localities.

References

Van Thong Nguyen. 1993. Initial evaluation of improving and regenerating measures in Cauhai Silvicultural and Experimentation Research Centre, Phu Tho Province (the essay of Master of forestry). Vietnam Forestry University.

Nhat Pham. 2001. Lectures on Biodiversity. Vietnam Forestry University.

Ngu Phuong Tran. 2000. Some Problems of Tropical Forests in Vietnam. Hanoi: Agricultural Publishing House.

Van Trung Thai. 1978. the Vietnam Flora. Hanoi: Scientific and Technical Publisher.

Van Trung Thai.1998. The Ecology of Tropical Forests in Vietnam. Hanoi: Scientific and Technical Publisher.

Ministry of Agriculture and Rural Development. Decision No.1565/QD-BNN-TCLN dated July 8, 2013 (the Ministry of Agriculture and Rural Development approving the forestry sector reform proposal).

15

The Role of Indigenous Communities in Rehabilitation of Degraded Forests in the Tropics

Prof. Shen Lixin[1] and Dr. Jaap Kuper[2]

Abstract

Indigenous communities could be perfect rehabilitators of degraded forest ecosystems. They live on the spot, have organisational structures, and they benefit from the rehabilitation, which is very important. The tragedy is that indigenous communities lack sufficient authority which is needed to govern over the surrounding natural resources.

Although forces from outside were generally the cause of the degradation and destruction, these outsiders were generally unable or unwilling to restore the forest after degradation. This resulted in vast areas of destroyed forest that poorly provide ecological services. It even threated traditional cultures and local knowledge on sustainable forest use. It left communities that were dependant of natural resources behind with reduced opportunities for development.

Some members of the communities still know how to use natural forest species, and which species, in order to restore forest in an ecological sound way. They are also able to pick up management skills that make badly degraded forest remnants develop into mature forests through community-tailored investment measures such as protection of natural regrowth, the release of natural tree saplings, small scale enrichment and selective afforestation. Small steps had immediate small success. It empowered the communities to improve their livelihood.

The community members which master the traditionally developed, ecological and local knowledge needed to use and restore forest in an ecological sound way are getting old.

---

1. Director of APFNet Kunming Training Center/Southwest Forestry University, China
2. International consultant on forest rehabilitation, former director of the Royal Forest Estate of the Netherlands

Therefore, local communities should get the authority and incentives needed to restore forest systems soon, before it is too late.

Background

The area of natural forest in the tropics is diminishing at a rapid pace. Much of it is being replaced by crops that quickly bring profits to a limited number of investors. At the same time, the area of wasteland is increasing. Much of the wasteland is the result of ongoing unsustainable timber harvest. The annual fires that often follow prevent secondary forest from developing, with the result that forests that are degraded or, in the worst case, destroyed, are unlikely to recover and so the wasteland remains.

The limited area that has been "restored to forest" has generally been transformed into tree plantations. We have tended to call these "forest" too, because they are composed of trees and produce timber. But in no way do plantations have the wide spectrum of values that natural forests have and in general they serve only the interests of individuals who have no links with local people.

There are, however, some examples where degraded or destroyed forest systems have been restored to as natural a forest as possible, but their area is very limited. This is surprising, considering that many communities in remote areas are very dependent on commodities from natural or semi-natural forests, and that millions of people who live in valleys are dependent on water resources that are regulated by forest-covered mountains.

Why are there so few such examples? Why is there so much wasteland? Typical arguments against restoring forest are:

- Tree planting is expensive and it takes too long to get returns on investments;
- Nobody takes responsibility for the land, so why invest in restoration;
- There's no point in planting trees because fire and livestock will prevent forest from developing.

If we accept these arguments then there is no hope that the wastelands will ever be converted into forests, nor is there any hope for the remaining natural forest: it will all be destroyed sooner or later. (For sure, it will be sooner).

Can something be done to counter these arguments? There are some possible solutions:

- Make planting cheap and with quick returns;

- Make people concerned about the land use so they will take responsibility for it;
- Ensure land management benefits those who take responsibility.

In this article I will show that tree planting can be cheap and beneficial, that people can become concerned about the land use and that land management can be well organised.

Make tree planting cheap

Traditionally, foresters think only of planting plantations. They plant huge numbers (3,000 to 50,000) saplings per ha so that the tree trunks will grow optimally for industrial timber products. After decades, such plantations produce large volumes of commercial timber. As well as being very expensive, the system is also vulnerable to fire and pests. Plantations are an investment-driven exercise. They are applied if prospects are good for returns on investment; if prospects are not good, plantations are not planted and the land is abandoned after the original forest has been cut. The sole objective of plantations is to generate returns on investment for the investors. The protection of water catchment, of forest resources for local communities or conservation of biodiversity have never been arguments for establishing plantations.

But meanwhile, it has become apparent that worldwide, water catchments, forest resources for communities and biodiversity are diminishing (Figure 1). So, who should take responsibility for these issues? And how?

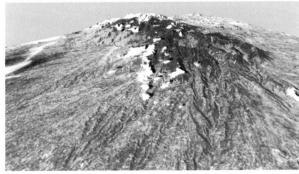

Figure 1 A typical example of natural forest diminishing and being replaced by wasteland

Notes: A typical example of natural forest (dark green) diminishing and being replaced by wasteland (light green and yellow). Unless action is taken, the system regulating water supply will disappear, and so will the forest commodities for local communities and the biodiversity.

Plantations are not the answer

Let us first consider how forest could be restored. It is unrealistic to expect much of the wasteland to be converted into plantations. And that is not even desirable! It would be much better to transform the wasteland into forest (natural or semi-natural). Compared with plantations, semi-natural forest is far more effective as a water catchment, as a resource providing a mix of forest products for local communities and as habitat for biodiversity. To guarantee permanent soil cover, the full spectrum of forest products and also the natural biodiversity, the forest should ideally be composed of a variety of indigenous tree species.

Just as importantly, to restore semi-natural forest it is not necessary to plant large numbers of saplings: 400 to 500 per ha is sufficient (see the section "Three levels of degradation"), and they could be seedlings or seeds obtained from remnant natural forests in the vicinity—for free! The seedlings or seeds collected should include the various species that compose the natural forest. They must be planted in a random mix of species, but taking care that the characteristics of the species match the conditions of the planting site (see the section "Three levels of degradation").

So, compared with conventional thinking in forestry, such restoration to semi-natural forest requires a completely different strategy that produces a completely different kind of forest.

Three levels of degradation

How degraded systems are converted into semi-natural forests depends on the level of degradation. It makes a great difference if some "forest climate" still remains on the site, and if there are wildings rather than completely destroyed forest such as pure grassland.

It is useful to recognise three levels of degradation.

- Totally destroyed forest: no trees remain;
- Severely degraded forest: some trees and forest climate remain, no wildings;
- Moderately degraded forest: some trees, forest climate and wildings occur.

Completely destroyed forest (Figure 2)

Strategy: plant entire area with mixed species in low numbers per ha

The starting position is no trees and therefore to create the new forest, saplings must be planted. They could be seedlings collected or raised from seeds from remnant natural forests somewhere in the vicinity. In order to increase their chances of survival after planting, the seedlings need to be grown on in a nursery (see the section "Nusery management"). Collected seeds must also be germinated and raised in the nursery.

Once the seedlings are fit for planting, they will be planted in lines in the field to make it easy to do the weeding during the first two to three years. The saplings are planted as mixed species. To reduce working time, the numbers are kept low: 400 to 500 saplings per ha is sufficient. The main purpose of the planting is to have sufficient numbers to create a forest and to shade out the competing grasses and weeds. Once the main competing vegetation has

Figure 2 Completely destroyed forest, showing erosion and landslide

been shaded out, the forest will develop by itself. Birds and mammals (notably fruit bats) will bring in new tree seeds from the natural forest.

It should be kept in mind that the microclimate in grassland is harsh. It is only fit for pioneer species that can tolerate the full light and heat. So, in such conditions the saplings must also include pioneer species, among them coppicing species that can soon be used as a source of firewood. Intermediate species can be added only in limited numbers. It is up to the birds and bats to bring in the tree species of subsequent succession stages once the grasses have been shaded out and some sort of forest climate has been created. Experience from Uganda has shown that this happens surprisingly quickly: only 5 years after the first saplings were planted, seedlings of climax tree species appeared spontaneously!

During the first 2–4 years and especially in the first year, competing vegetation must be weeded out. The traditional way is to slash the competing vegetation, but trampling it has also proved efficient. The most important way to reduce competitive species that have rhizomes is repeated weeding, to weaken the rhizomes. This means investing extra effort in the first 6 months after planting the saplings. If the competing vegetation is given time to overgrow the saplings, it will hamper their growth and rhizomes of competing plants will build up their reserves, making them more difficult to eradicate. This must be prevented, to avoid a setback in the coming years.

Summing up, the actions needed are:

- To collect seedlings or seeds from nearby forest;
- To grow them on or raise them in a nursery;
- To plant a mix of sapling species;
- To weed for 2–4 years.

Severely degraded forest **(Figure 3)**

Strategy: enrichment planting.

Figure 3 Severely degraded forest showing climber carpet

In this case, some trees and forest climate remain after the logging. Because of the sudden increase in light, climbers often become dominant and cover the entire area. Under such conditions, the development of the forest is hampered for a long period. Most of the tree species that were at home in this site have been cut, so they cannot supply seed. Even if some seedlings of these trees are present, they will not survive competition from the climbers. The problem can be solved by cutting the climbers for 2 consecutive years and planting saplings of intermediate and climax tree species in the "half-shade" of the sparse older trees. The saplings should have been collected as seedlings or seeds and grown on or raised in a nursery, as described above. In enrichment planting too, a maximum of 400 to 500 saplings per ha planted in lines suffices.

Summing up, the actions needed are:

- To collect seedlings or seeds from nearby forest;
- To grow them on or raise them in a nursery;
- To plant a mix of sapling species as enrichment in the degraded forest;
- To weed for 2–3 years.

Moderately degraded forest

Strategy: release wildings

Some trees remain, creating a forest climate. Wildings of intermediate and climax tree species are present. However, similar to the previous situation, the development of the forest is blocked by an abundance of climbers, and sometimes by less desirable pioneer trees. It is sufficient to cut the climbers and some of the pioneer trees. Again, a maximum of 400 to 500 saplings per ha need to be released from competitive vegetation.

Summing up, the actions needed are:

- To find saplings of desired tree species in the degraded forest;

- To release the saplings from climbers and less desirable competitors during 2–3 years.

Nursery management

The seedlings or seeds collected in forest remnants need to be grown on or raised in a nursery until they have grown 30 to 50cm tall and have developed a good root system. The nursery may be simple and small. Many farmers are capable of setting up such a nursery. The seedlings should be planted in polyethylene bags containing soil that is mainly forest soil, because in order to grow properly, the seedlings need the beneficial soil fungi (mycorrhizae). Regulate the shade in the nursery to mimic the conditions in the forest from which the seedlings come from and gradually accustom them to the conditions at the site where they will be planted later on. Planting must be done at the start of the wet season.

Make people concerned about the land use

Who will do this forest restoration? And who will protect and manage the restored forest?

The forest could, of course, be restored by the authorities that are responsible for assuring water supply and protecting biodiversity. But so far they have not done so, perhaps because they have other priorities. Even if they were to restore the forest, that would not solve the question of who is responsible for the land use. For this, it is worth looking to the local communities. After all, they live on the spot and will benefit most from the ecologically restored forest.

Let us look at the needs of local communities. These are firewood, building materials, fodder, grass and many other non-timber forest products. That is precisely the product mix that can be taken from restored natural or semi-natural forests.

Would local communities be capable of restoring and managing the forest? Certainly! Their strength is that they can literally oversee their forest: they are aware of its benefits, they can spread the restoration activities over longer periods, and they welcome the small returns that the forest brings them as individuals. Very importantly, once they have invested in restoration, they will organise themselves in order to protect and manage their restored natural resource.

Deadlocks to be broken

Why do not local communities engage in massive forest restoration? Is it because they are unaware of how relatively easy and cheaply it may be done? They may have been reluctant to act because they believe that forest restoration can be done only by creating (unwanted) plantations. This could be solved by creating demonstration plots. But someone needs to take the initiative to set these up!

A more intractable problem is the absence of land titles. Much wasteland is neither communally nor privately owned.

Much wasteland is state-owned in one way or another and therefore individuals or communities are reluctant to risk investing in such land because of the uncertainty about whether they will later be able to reap the rewards of their investment. It is therefore necessary for communities to get in touch with the authorities so that formal agreements can be made about the long-term usufruct of the land on which communities are investing in forest restoration. Such agreements should be possible if it is acknowledged that communities and authorities have compatible interests in forest restoration: the communities can have the mix of forest commodities produced and the authorities want to realize erosion control, water regulation and economic development. Both parties benefit from mountain slopes that are permanently covered by sustainably managed forest.

Who will take the first steps?

It is clear from the above that once they have been trained in ecological restoration, the most logical parties to involve in ecological forest restoration are local communities. However, they are seldom in a position to start the restoration. The most frequent obstacle is the absence of land titles. Local communities are rarely considered to be serious candidates for negotiating these with the authorities. On the other hand, the authorities do not take the initiative either. They do not recognise the opportunities and probably have other priorities.

Conclusions

Ecological forest restoration is easy to apply. It can be done at low cost and by local communities.

Not only may the ecologically restored forest be sustainably used by local communities, it also provides restored water catchment and habitat for the indigenous flora and fauna.

To start ecological forest restoration on wasteland, it is necessary to make long-term agreements with the landowner because of orest restoration benefits everyone.